INTERNATIONAL CASE STUDIES IN TOURISM MARKETING

This international case study book provides 27 expertly curated case studies on the topic of tourism marketing, each with detailed implementation instructions for the instructor in order to maximise student participation and learning.

The dynamic characteristic of the industry under the influence of micro and macro environment factors requires future professionals to be equipped with appropriate skills and competencies to deal with such factors in real-life practices. Curated and developed by industry experts and practitioners, these case studies embody imaginary and real-world scenarios with the aim of best preparing students for their future careers. This compelling set of case studies follows a logical and uniform structure and covers topics such as marketing mix, crisis management, digital marketing, quality development, product development and sustainability.

With reflective questions throughout to aid both in-class discussion and self-study, this book is an ideal study resource for use in higher and vocational education, and its unique, teaching-led approach positions it as a vital study tool for instructors and students alike.

Gürhan Aktaş is Associate Professor at the Department of Tourism Management, Faculty of Business, Dokuz Eylul University, Turkey. He holds a BA in Business Administration from Hacettepe University, Turkey, an MSc in Tourism Management from the University of Surrey, UK and a Ph.D. in Tourism Marketing from Bournemouth University, UK. He delivers both undergraduate and postgraduate courses in the fields of destination management, tourism geography and tourism marketing. He has authored academic publications on crisis management in tourist destinations, visitor attractions and events management.

Metin Kozak holds a Ph.D. in Tourism from Sheffield Hallam University, UK. He has contributed a wide range of articles to top-tier journals and conference papers in more than 40 countries, and over 30 books released by international publishers. He has been involved in several national and international research projects, particularly with his partners based in Europe, Asia and the US. He received the EFQM Ph.D. Thesis Award and a number of conference paper awards. He acts as the co-editor of *Anatolia* and is a member of the editorial/review board for many international journals. His research interests entail marketing and consumer behaviour in an interdisciplinary context. He is currently affiliated with the Faculty of Communication at Kadir Has University, Istanbul, Turkey.

ROUTLEDGE INTERNATIONAL CASE STUDIES IN TOURISM

Edited by **Gürhan Aktaş**, *Dokuz Eylul University, Turkey* and **Metin Kozak**, *Kadir Has University, Turkey*

International Case Studies in Tourism Marketing
Edited by Gürhan Aktaş and Metin Kozak

For more information about this series, please visit: www.routledge.com/Routledge-International-Case-Studies-in-Tourism/book-series/ICS

INTERNATIONAL CASE STUDIES IN TOURISM MARKETING

Edited by Gürhan Aktaş and Metin Kozak

Routledge
Taylor & Francis Group
LONDON AND NEW YORK

Cover image: © Getty Images

First published 2022
by Routledge
4 Park Square, Milton Park, Abingdon, Oxon OX14 4RN

and by Routledge
605 Third Avenue, New York, NY 10158

Routledge is an imprint of the Taylor & Francis Group, an informa business

British Library Cataloguing-in-Publication Data
A catalogue record for this book is available from the British Library

Library of Congress Cataloging-in-Publication Data
A catalog record has been requested for this book

ISBN: 978-1-032-02315-1 (hbk)
ISBN: 978-1-032-02313-7 (pbk)
ISBN: 978-1-003-18285-6 (ebk)

DOI: 10.4324/9781003182856

Typeset in Bembo
by Newgen Publishing UK

CONTENTS

FIGURES

TABLES

ABOUT THE CONTRIBUTORS

Vedat Acar is currently an Assistant Professor at the Department of Tourism Guiding, School of Tourism, Aydin Adnan Menderes University, Turkey. He has been a licensed tourist guide of Turkish Ministry of Culture and Tourism since 2009. Prior to joining the university as a research assistant in 2012, he took part in guided tours in Turkey for several years. Since 2016, he has been conducting research in the field of tourist behaviour on guided tours in Turkey. He is the co-editor of the Journal of Qualitative Tourist Guiding Research.

Maria D. Alvarez is Professor of Tourism Marketing at the Department of Tourism Administration at Boğaziçi University, Turkey. She has co-edited books for prominent international publishers, and has published in leading academic journals. She has also consulted on and been involved in both locally and internationally financed destination development projects. Her research interests include sustainable destination development, destination marketing and tourist behaviour.

Luisa Andreu is Associate Professor of Marketing at the University of Valencia, Spain. She has been a visiting scholar at Penn State University, USA (2008), University of Cambridge, UK (2011) and University of Surrey, UK (2019). Her main research interests are tourism destinations, sustainable marketing and consumer behaviour in the tourism and hospitality industry. Her research on tourism has been published in *Annals of Tourism Research*, *Current Issues in Tourism*, *International Journal of Hospitality Management*, *Journal of Sustainable Tourism* and *Tourism Management*, among others. She is the coordinator of the Double Degree in Business Management and Tourism at the University of Valencia, Spain.

Andrés Artal-Tur is an Associate Professor at the Department of Economics at Technical University of Cartagena, Spain. His main research interests focus on tourism economics, tourist behaviour with applications in the field of tourism marketing and destination planning, culture and tourism and tourism sustainability. He has extensively published his research findings in international impact journals, and is the editor of books for well-known publishers such as Springer, Routledge, Elsevier, Emerald and CAB International.

Monika Barnwal is a researcher and Ph.D. candidate at the Department of Tourism Management, Jamia Millia Islamia Central University, New Delhi, India. Her main line of research focuses on smart tourism destination and destination branding. She holds a bachelor's degree in Computer Science and an MBA in Tourism Management and has published seven Scopus-listed journals and National listed UGC-CARE Journals. She has authored one tourism book on competitive exams. She maintains her blog, as well as her YouTube channel, on a regular basis where she engages with different segments of tourism.

Enrique Bigné is Professor of Marketing at the University of Valencia (2001–) and was formerly at Jaume I University, Spain (1996–2001). He has been a visiting scholar at the University of Maryland (2011, 2012) and Berkeley Haas School of Business, USA (2014). His main research interests are tourism destinations, digital communication and consumer neuroscience. His research on tourism has been published in *Annals of Tourism Research*, *Current Issues in Tourism*, *Journal of Travel Research* and *Tourism Management*, among others. He has served as Head of Department, Vice-Dean, Dean and Vice-Chancellor. He is the Editor of *European Journal of Management & Business Economics*.

Erhan Bilgici graduated from the Department of Tourism and Hotel Administration at Akdeniz University, Turkey. His 24 years of experience in the tourism and hospitality industry include positions in hotels such as Rixos Premium Tekirova Villas Sales and Operations Director; Rixos Sungate F&B Manager; Adam and Eva Hotel; Kempinski Hotel Mall of Emirates Dubai, UAE; Kempinski Hotel the Dome Belek, Turkey; Kanuga Conferences Centre, USA. He is currently the Assistant General Manager at Rixos Sungate Hotel, Antalya Beldibi, Turkey.

Monika Borowiec-Gabryś is currently Assistant Professor at the Department of Entrepreneurship and Spatial Management in the Institute of Geography at the Pedagogical University of Cracow, Poland. She holds a Ph.D. in Natural Sciences (Geography). Her main line of research focuses on knowledge-based economy, the role of higher education and academic centers in processes of socio-economic transformation, entrepreneurship education, Christian ethics, processes of globalisation and European integration.

Demet Ceylan holds an MSc in International Tourism Management from Akdeniz University, and a BA in Business Administration from Boğaziçi University, both in Turkey. Her 25 years of industry experience includes Financial Controller, CFO and Member of the Board at multinational tourism and hospitality businesses. She is a Ph.D. candidate and provides consultancy services to hotel chains, ABU Tourism Research Centre and Antalya City Council. She actively lectures at Antalya Bilim University, Turkey and runs a 'Management Shadowing Program' providing a platform for the sector and prospective graduates at managerial level.

Evi Chatzopoulou holds a Ph.D. in Marketing from University of Piraeus, Greece. She is adjunct lecturer in Consumer Behaviour at University of Patras, Greece specialising in the area of tourism marketing. She has published articles in academic journals and she has presented scientific papers in international marketing conferences. Her main line of research focuses on consumer behaviour, with an emphasis on consumer psychology, tourist behaviour and consumer behaviour in crises.

Carol Yi Cui is an experienced digital marketer, specialised in the tourism industry. She received her bachelor's degree in Tourism Management from School of Hotel and Tourism Management, The Hong Kong Polytechnic University, SAR China. Her track records in social media include Hong Kong Tourism Board (DiscoverHongKong.com), Gansu China and Visit Melbourne. She developed Onesight.com, a social media marketing management platform offering consultancy services to brands.

Tevfik Demirçiftci received his M.Sc. in Hospitality Information Management from the University of Delaware and his Ph.D. in Hospitality Management from the University of Nevada Las Vegas, both in the USA. He is an Assistant Professor at East Stroudsburg University, USA. He is a certified Excel Specialist and Expert approved by Microsoft Office. He has 22 years of experience in casinos (MGM-Borgata Hotel Casino and Spa), hotels (Marriott, Ritz-Carlton) and restaurants (Buddakan Atlantic City) in Bahrain, USA and Turkey.

Maurizio Droli is currently a Senior Research Fellow at the Department of Environmental Sciences, University of Udine, Italy. As a sociologist holding an EU Ph.D. in Resource-based Economics, he conducts research in ecosystem service innovations, social innovation and green-based health strategy definition. Since 2012, he has authored approximately 60 scientific contributions. He also helps general medicine practitioners to prescribe evidence-based outdoor activities to their patients.

Vahid Ghasemi is Assistant Professor at Universidade Europeia, Portugal and integrated research member in CEFAGE, Évora, Portugal. His research areas include marketing, destination management, luxury tourism and tourist behaviour. He has published papers in several tourism and hospitality journals such as *Journal of Destination Marketing and Management*, *Tourism Analysis*, *Anatolia* and *European Journal of Tourism Research*, among others. He received the Emerald Literati Award 2020 for Outstanding Author Contribution.

Su Gibson currently serves as the Assistant Director for University of South Carolina Beaufort's Center for Event Management and Hospitality Training, USA. She is also an instructor for the Hospitality Management Department at the University of South Carolina Beaufort. She has more than 20 years of industry experience in the hospitality industry, including traveling nationally and internationally as an inspector for the Forbes Travel Guide. Her research interests include service quality evaluation and management, and the emotions, setting, language and actions related to feelings of being welcome or unwelcome in a service environment.

Rut Gomez Sobrino is currently teaching at Shanghai International Studies University, China. She graduated from Communications at University San Pablo CEU and holds an MA in International Relations from Istanbul Bilgi University, Turkey and in Euromediterranean Studies from Cairo University, Egypt. She also completed a Postgraduate Diploma in Development Management and postgraduate certificates in Gender and Social Development and Poverty Reduction at the University of London, UK. Rut has worked under the UN framework since 2007, having the opportunity to serve in Africa, the Middle East, Asia and Latin America for over a decade. Until present times, she worked for six UN agencies (UNESCO, UNDP, UN Women, UNCDF, UNWTO and UNEP) as well as IUCN. Under UNWTO (2015–2018) she conducted training for over 100 governments on crisis management of tourism destinations. In 2021 she created a programme on gender equality policies in tourism for UNWTO Academy.

H. Kader Şanlıöz-Özgen is affiliated as a full-time Assistant Professor at the Department of Hotel Management, School of Applied Sciences, Özyeğin University, Turkey. She holds a Ph.D. in Tourism Management from Dokuz Eylul University, Turkey. Her field of expertise is in the areas of hospitality marketing, hospitality experience and quality management.

Lanlan Huang is a Senior Lecturer and Researcher at the Shanghai Urban Construction Vocational College, China. She is a Ph.D. Candidate at the School of Hotel and Tourism Management at the Hong Kong Polytechnic University, SAR China. Her research interests include vocational education in hotel and tourism management, B&B tourism, rural tourism and host-guest interaction.

Aise Kim is a Senior Lecturer in the School of Management at the University of South Australia. Her teaching and research interests include tourist behaviour, tourism marketing, destination management, ecotourism, food and wine tourism and sustainable tourism. She actively engages in working with the tourism industry at both local and international levels. She serves as a regional editor of *Anatolia* and an executive committee member of Wildlife Tourism Australia Association, Australia.

Özen Kırant Yozcu is currently a visiting professor in ISAG (European Business School) in Porto, Portugal and part time instructor at Boğaziçi University and Bilgi

University, both in Turkey. Previously, she was a Sales and Marketing Manager in the hotel industry. She also consults and is involved in many local and international corporate projects. Her research interests include hotel management, marketing and marketing communication management, luxury management and event management. She is a board member and Director of Education of Meeting Professionals International (MPI) Turkey Club.

Burçin Kırlar-Can is currently an Assistant Professor in Recreation Management at the School of Tourism, Pamukkale University, Turkey. She holds a BA in Tourist Guiding and an MA in Turkish Art from Ege University, Turkey, as well as an MA in Tourism Management from Dokuz Eylul University, Turkey. She also holds a Ph.D. in Tourism Management from Dokuz Eylul University, Turkey. Her research areas focus on risk perception, crisis management and tourist behaviour.

Maximiliano E. Korstanje is Senior Lecturer at the University of Palermo, Buenos Aires, Argentina. He is the book series editor of *Advances in Hospitality, Tourism and Service Industries* (IGI Global, US) and *Tourism Security Safety and Post Conflict Destinations* (Emerald Group, UK).

Salar Kuhzady is Assistant Professor at the Department of Tourism Management at the University of Kurdistan, Sanandaj, Iran. His main research interests include e-tourism, smart tourism and digital enterprises. In addition to international academic experiences in Turkey and Portugal, and publishing 30 academic articles, he has working experiences in different tourism start-ups.

Vijay Kumar has over 15 years' teaching experience and is currently working as an Assistant Professor at the Department of Tourism & Hospitality Management, Jamia Millia Islamia Central University, New Delhi, India. He has approximately 30 research papers and other publications, including books, to his credit.

Drew Martin is Professor of Tourism and Hospitality Marketing at University of South Carolina, Turkey. Formerly, he was Professor of Marketing at University of Hawaii at Hilo, USA. His research examines advertising, services marketing, consumer/visitor behaviour and comparative cultural studies. He has served as Senior Associate Editor of *Buyer Behaviour, Journal of Business Research* (2012–2015), Associate Editor for the *International Journal of Culture, Tourism and Hospitality Research* (2007–2013) and on boards for the Hawaii Small Business Development Center (2016–2017) and South Carolina Restaurant and Lodging Association, USA (2017–2020).

Juan Pedro Mellinas is Assistant Professor of Marketing at University of Murcia, Spain. His research interests are focused on tourism satisfaction, using online reviews in websites like Booking.com or TripAdvisor as a main data source. He has published more than 30 academic articles and conference proceedings in journals like *Tourism Management*, *Annals of Tourism Research*, *International Journal of Hospitality Management* and *Tourism Review*, among others.

Peter O'Connor is Professor of Strategic Management at the University of South Australia. His primary research and teaching interests focus on the effect of digital on business, particularly on retailing and marketing. Peter has published in leading academic journals including the *Journal of Marketing*, *Harvard Business Review*, *Journal of Retailing and Consumer Services*, *Tourism Management*, the *Cornell Quarterly* and the *International Journal of Hospitality Management*, among others. Prior to joining UniSA, Peter founded the Chair in Digital Disruption at Essec Business School in France, where he held a variety of other academic roles.

Yasuo Ohe is Professor at the Department of Agribusiness Management, Tokyo University of Agriculture, Japan. His main research involves economic analyses of rural tourism and farm diversification to establish a viable rural business in tourism and hospitality. He is currently Chair of the project evaluation committee on rural tourism in the Japanese Ministry of Agriculture, Forestry and Fishery (MAFF) and President of the Japan Society for Interdisciplinary Tourism Studies. He received the Sohn Hai-Sik Award and Best Paper Award from APTA, and the Distinguished Service Award for Agricultural Technology from the Japan Agriculture, Forestry and Fisheries Research Council. His recent publication is on community-based rural tourism and entrepreneurship.

Marcelo G. Oliveira is Associate Professor and Coordinator of the Area of Tourism and Hospitality at Universidade Europeia's Faculty of Social Sciences and Technology, Lisbon, Portugal. He is a researcher of the Centre for Lusophone and European Cultures of the University of Lisbon, and is a regular contributor to international publications and is responsible for the seminar on 'Case Studies in Tourism and Hospitality' at Universidade Europeia's Master's in Tourism Management Programme.

Serim Paker is a graduate of the Maritime Transportation and Management Engineering Program, Istanbul Technical University, Turkey. He worked for dry bulk cargo ships for nearly eight years prior to joining the Faculty of Maritime, Dokuz Eylul University, Turkey. His Ph.D. is on the marketing of marine tourism and his research interests include service marketing, marine tourism, and underwater and nautical archaeologies. He directs two postgraduate programmes in marine tourism and has been a member of the 360 Degrees Historical Research Group for 16 years, serving as a vice president for six years, and performing experimental archaeology research projects on naval architecture, naval history and ancient navigation.

Andreas Papatheodorou is currently a Professor in Industrial and Spatial Economics with Emphasis on Tourism at the University of the Aegean, Greece, where he also directs the MSc Programme in Strategic Management of Tourism Destinations and Hospitality Enterprises and the Laboratory for Tourism Research and Studies. He also holds appointments as Adjunct Professor at the University of New South Wales, Australia and Frankfurt University of Applied Sciences, Germany. He is the President of the Hellenic Aviation Society and a Member of the Executive Board of the Hellenic Civil Aviation Authority.

Ioulia Poulaki is currently an Assistant Professor in Tourism Business Administration at the University of Patras, Greece, where she is also a member of the Laboratory of Information Systems and Forecasts in Tourism (TourISFlab) of the Department of Tourism Management. Her research interests focus on the fields of air transport management, tourism distribution channels, digital tourism and tourism development, while her papers have been published in international journals, collective volumes and international conference proceedings. She has also ten years of experience in the air transport industry, with high specialisation in revenue management and distribution systems of airline services.

Noelia Sánchez-Casado is Associate Professor of Marketing at Universidad Politécnica de Cartagena, Spain. She has a BA and Ph.d. in Business Administration, and a MSc in Tourism Management. She has taken part in international marketing conferences, including the Association for Consumer Research Conference and the European Marketing Academy Conference, and she has published articles in international journals and books. Her research is focused on the use of social media as a marketing tool, and relationship and tourism marketing.

Juan Sánchez-Villar holds a degree in Business Administration, a degree in Political Science and a Ph.D. in Marketing. His professional experience is focused on the tourism and hospitality industry. He is partner and COO of Caro Hotel and Alma del Temple restaurant, as well as founding partner and CEO of Calma Ventures, a company dedicated to the development of unique tourism projects such as Villa Indiano or Convent Carmen. Juan is an Associate Professor at the University of Valencia, Spain, where he teaches tourism marketing, and also collaborates in postgraduate training programmes within the areas of strategy and business development.

Patricia Isabel Santateresa-Bernat is currently a Professor in the Business Department at Florida Universitaria, and at the Valencian International University, both in Valencia, Spain. Her main line of research focuses on destination image, with a special emphasis on influencer marketing, and also on the distribution and marketing of video games. She has extensive experience in the field of educational innovation, especially in the application of ICT.

Hugues Séraphin is a Senior Lecturer in Event/Tourism Management Studies and Marketing. He holds a Ph.D. from the Université de Perpignan Via Domitia, France and joined the Business School at The University of Winchester, UK in 2012.

Abdullah Tanrısevdi is currently a Professor in Tourism Marketing, Research Methods and Special Interest Tourism at the Department of Tourism Guiding, Faculty of Tourism, Aydın Adnan Menderes University, Turkey. He visited Oklahoma State University – Centre for Hospitality and Tourism Research, USA, as a visiting scholar (2011), and received a post-doc from the University of South Florida, USA (2015). He is the Editor-in-Chief of the Adnan Menderes University *Journal of Travel and Tourism Research.*

Małgorzata Zdon-Korzeniowska is currently Assistant Professor at the Department of Entrepreneurship and Spatial Management in the Institute of Geography at the Pedagogical University of Cracow, Poland. She holds a Ph.D. in Economics in the discipline of management. Her research interests include market orientation of local governments, territorial marketing, management of tourism at local and regional level, and regional tourism products.

Elaine Yulan Zhang is a Lecturer at the Sino-British College, University of Shanghai for Science and Technology, China. She received her Ph.D. in Tourism Management from the School of Hotel and Tourism Management, The Hong Kong Polytechnic University, SAR China. Elaine previously worked as E-marketing Officer at China Travel Service (H.K.) Limited, responsible for online marketing. Her research interests are luxury travel, tourism marketing and responsible business.

Alejandra Zuccoli is Professor of Tourism Marketing at the University of Palermo and the University of Salvador, both in Argentina. She is the director of the Joy Labs at the University of Palermo, Argentina. She was distinguished as an innovator of method ALTAX to improve human hospitality applying neuromarketing to education.

PREFACE

The *Routledge International Case Studies in Tourism* series is projected to fill in a gap in the existing textbook collections on tourism and its related subject fields. Most of the existing textbooks in the field contain cases mostly as end-of-chapter sections and mainly embrace cases with in-class discussion questions listed in the end. Therefore, the execution of such case studies does not go beyond distributing or sharing case-related information with students, and encouraging them to discuss case-related issues individually or as a group. Instructors are often not provided with enough information on how a case study should be implemented, what different implementation methods could be used and what outcomes are expected from the application of the case.

Such mundane structure of cases often leads both instructors and students to lose enthusiasm in in-class applications. Moreover, cases on specific subfields and headings are often limited, with a few included in a number of tourism textbooks, requiring instructors and students to search a variety of sources to find appropriate cases matching their course content. This project, therefore, aims to combine a number of cases on specific tourism-related subjects in book series, each of which could, then, serve as a key reference of cases on the concerned subjects.

The series is envisioned to embrace cases on various tourism subjects, expansively found in the curriculums of higher education programmes. Following this first volume on tourism marketing, the series is planned to continue with other tourism-related subjects. Each book in the series will gather cases on one of these specific subjects and will provide opportunities for students to apply their theoretical knowledge into real life cases or imaginary scenarios. The difference of the book from existing case study material is that each case study is presented in such a way to include: the aim and objectives of the case study; the expected learning outcomes from its application; the required background knowledge of students for its effective implementation; the steps of implementation in class (or online); and further reading and research suggestions. This structure is designed to make the use of cases easier for instructors and trainers.

Case Studies in Tourism Marketing

The series starts with the first volume on tourism marketing. When selecting cases for the volume, the following criteria has been used: the geographic distribution of cases representing tourism destinations and tourism, travel and hospitality establishments across the world; the probable achievements of a case with regards to its learning outcome objectives; and creative and innovative cases embracing alternative teaching techniques and solutions –i.e. role plays, brainstorming, hands-on practice, in-class model designs and company/destination visits.

Each case follows a pre-determined format, starting with the aim/objectives of a case, and concluding with further reading and research suggestions. Since the aim of the series is to encourage students to engage in research to understand and evaluate the concepts covered in each case, the background theoretical knowledge is not presented in detail, but only briefly. As a result, most case studies are short in length, but with some tables and graphs for students to work on. The only exception to this structure is found in Case 1. This case represents the findings of a project, which was designed to understand the impact of pleasurable experiences in education. The case is included in this volume as an example of a project on the use of different teaching methods in tourism education, and those readers interested in repeating similar studies are advised to contact the authors.

As the first volume of the *Routledge International Cases in Tourism* series, the book aims to provide instructors and students with a number of cases on various topics of tourism marketing. The range of marketing topics covered is extensive and includes market segmentation, pricing, communication strategies, product development and media relations. With such a range of cases on tourism marketing, it is hoped the book will become an essential supplement to textbook material. While the instructors will have an opportunity to enrich their lectures with cases entailing alternative teaching methods, students will be able to comprehend the applicability of theoretical knowledge to real life events. We believe that the use of tourism cases in higher education is especially crucial in preparing students for their future careers. The dynamic characteristic of the industry under the influence of micro and macro environmental factors requires future professionals to be equipped with appropriate skills and competencies to become familiar with and to deal with such factors.

Content and Target

The cases are written by the contribution of 41 authors, who have either academic or professional expertise in tourism marketing. The book brings together 27 cases, each looking into the different components and stakeholders of the tourism system. Hence, readers will be able to find cases on tourist destinations, hotels, travel agents, restaurants and airlines, as well as on various topics including segmentation, positioning, marketing mix, social media, digital marketing, crisis management, quality development, loyalty, education, revenue management, market/ing research, image, innovation, product development and sustainability. The cases include both success and failure stories.

This series is targeted at both instructors and students in a broad range of tourism fields in higher education (both undergraduate and postgraduate), but it can be extended to those studying/teaching other close fields such as business, economics, marketing, finance, accounting and communication. The case studies will provide additional material to support textbooks or other materials used for teaching in the classroom, as well as for homework and exams.

Although it would be impossible to cover all sub-topics of tourism marketing in a single case book, the number of cases is determined according to the duration of academic semesters in higher education programmes across the world – i.e. ranging anywhere between 12 to 15 weeks for a semester. In other words, the instructors and students, with 27 cases included, will be able to have case study material on tourism marketing at least for one academic semester. In fact, when combined with theoretical lectures and presentations, the book could also be used for the whole academic year.

The cases vary in length between 1900 and 5750 words depending upon the content. The geographical distribution of contributors is quite diverse, representing 14 countries across the world: Argentina, Australia, China, France, Greece, India, Iran, Japan, Poland, Portugal, Spain, Turkey and the USA. Since the authors have been encouraged to design their real-life or imaginary case studies to enable users to understand the concepts, theories and managerial applications of the selected themes, the series also has the potential to be a good source for vocational and on-the-job training programmes. Therefore, any type of tourism establishment, ranging from hotels to visitor attractions, will be able to use the series in training their employees.

Although cases comply with academic writing rules, they are not as detailed as journal articles or book chapters. They only briefly mention the key theories and concepts related to each case, and direct readers to other publications for further essential reading. We believe that most academics would find the series extremely helpful as a teaching resourse, and would like to add these collections to their libraries. As such, we hope that the series will accomplish its stated objectives in the near future.

Gürhan Aktaş
Dokuz Eylul University, Turkey
Metin Kozak
Kadir Has University, Turkey

Case 1

THE JOY LABS

PANCOE as a new instrument to enhance tourism education

Alejandra Zuccoli, Hugues Séraphin and Maximiliano E. Korstanje

Duration

A total of 18 encounters lasting 90 minutes each.

Learning objectives

Upon completing the case, participants will be able to:

- facilitate the inclusion of foreign students who are joining a course as year 1 students in tourism and hospitality
- optimise the student learning experience, alongside reducing the rate of student drop-out
- implement innovative methods to improve traditional tourism education, using tourism marketing case studies.

Target audience

The present case is oriented to stimulate the learning skills of pre-graduate students in the first year of a tourism bachelor's degree. Methodologically speaking, students studying a BA in hospitality were excluded from the experiment. The study started to meet the needs of new students coming from different socio-economic and national contexts to the first years of a tourism BA in Argentina. Pre-graduate students (participants), who were involved in this experiment, came from different Latin American countries such as Colombia, Ecuador, Venezuela, Brazil and Chile (only to name a few). In all cases, they did not reside with their families in Argentina. They arrived only to study tourism with the intention of returning to their countries. It is safe to say that these students faced an unknown landscape and a climate

DOI: 10.4324/9781003182856-1

of emotional deprivation. Hence, these students were low-rated in the preliminary exams in comparison with Argentinean students.

The main goal of the laboratory was to associate the creativity of gifted students with artificial intelligence and digital social networks such as Twitter. Secondly, the Laboratory of Joy explored and acknowledged the importance of smart ideas generated through the formation of safe social networks. A third goal was address the need to reduce university drop-out rates. The students' mingling allowed rapid optimisation of learning. Students took direct intervention in diverse team-building activities like cooking, food tasting and baking bread.

The case study explores the role of pleasure (joy) as a catalyst for new emerging learning abilities and skills to potentiate students' performance in tourism and education marketing.

Teaching methods, sampling and equipment

The experiment was run using two different focus groups (10–20 participants in each group). The experiment was based on a sample of 870 participants who regularly use Twitter (Twitter-@holapancoe). The sample was divided into active and passive participants. Whilst the former refers to tourism undergraduate students who took part in PANCOE (40 students), the latter refers to students coming from other universities or areas, but who were not involved in the PANCOE (830 participants). The cohort ranged from 18 to 25 years old. Initially, the sample was made of tourism undergraduate students taking a specific course called Integracion, Ambientacion y communication (Communication, environmentalism and integration). PANCOE looked at integrating the senses of students with their emotions, academic performance and the digital platforms. PANCOE devoted efforts to transform negative feelings like fear into positive ones like joy. The experiment used different digital platforms like Twitter and Facebook, as well as culinary elements, such as baking bread and cooking.

Case

It is important to mention that the experiment was divided into three clear-cut steps. Each one was performed differently but they are interrelated over the year.

Step 1

Participants of this experiment were instructed to form two groups. Each one had 10–20 participants. PANCOE involved 18 encounters at the School of Tourism, the University of Palermo, Argentina. Each encounter lasted about 90 minutes. The experiment was divided into three facets where students developed different skills and learning processes. Professor Alejandra Zuccoli introduced the students to the importance of engaging in and experiencing new and innovative education techniques. For approximately 10 minutes, each group was exposed to positive stimuli, such as smelling different fragrances and perfumes. Over the next 20 minutes, students were bombarded with pictures containing paradisiacal landscapes and beaches. They were asked to remember their best trip as well as the emotions and experience such a trip evoked. They were encouraged to travel in their mind without moving. Imagination played a leading role in this process. Students were encouraged to give the keywords to describe such an experience. At the third stage, students tasted different regional dishes and drinks, which came from their home countries. Joy Lab Director Alejandra Zuccoli started from the premise that these pleasurable moments, which were repeated in this order during each meeting, released endorphins whilst also improving students' attention and performance in the classroom. The experiment took place on different days of the elective year, bolstering a fluid dialogue between professors and students. It was vital at this stage that each group worked in isolation and members had no previous communication about their experiences. The first group, group A, was directly involved with the PANCOE experiment, whilst the second group, group B, was enrolled in the traditional programmes and syllabuses of the tourism school.

Step 2

Students in group A were invited to bake bread as well as cooking different traditional dishes. Later they were encouraged to share their products with their peers. They were invited to take pictures of the dishes and pieces of bread. Each participant created a Twitter account and published the pictures, inviting 50–60 external followers. They shared the information, dishes, limitations and feelings in their Twitter account. This process took about 50–60 minutes and helped students to communicate their expectations with members of group B. This ignited a much deeper interaction which was based on the needs of selling their products and increasing their number of followers, whilst positioning their products to gain further traction. Students received constructive feedback from followers whilst they overtly manifested not only their expectations but also hopes and fears during the experience. Each student provided its commentaries in writing (anonymously, in a letter left in a ballot box). Each letter was carefully read and reviewed by Professor Zuccoli.

Step 3

One day of the year, a final exposition was organised for all students to share their pieces of bread, stories and dishes with other students who have not taken direct participation in the experiment. In this event, each student had a particular booth. These booths were visited by attendants, as well as important Argentinean Chefs who tasted the dishes, whilst sharing their experience and giving fruitful instructions to students. At this stage, the experiment not only re-integrated multiple sensory dimensions in a friendly climate of cooperation, but also settled in students the feeling they were part of something really important. Once the event ended, Professor Zuccoli convened a new encounter – a final exam. Students were evaluated and subjected to regular tests and verbal and written exams. Each exam lasted 2 hours and the results were ultimately communicated within seven days. Per rules and procedures of the University, the rigor of exam and the content was the same for students who had participated in PANCOE and for those who had not. Students were evaluated in the respective exams by different professors who taught their courses. The professors reviewed each exam using the double-bind method (professors were not aware of whether the student was involved in the experiment or not). At the same time, authorities closely followed up each student`s performance as well as their grades and the evolution of their careers. Finally, a drop-out rate was created to follow up how many students who took part in PANCOE abandoned their studies. These exams, which took place in six elective years, included classic curriculum content.

Results and discussion

The results matched expectations as the students involved in the programme performed better than the others (the capacity of memory storage increased notably to 33% in students who took part in PANCOE, in comparison to students who did not participate in the experiment, who had a memory storage capacity of 20%). The percentage of students who graduated was 40% amongst those involved in the programme compared to 28% for the others. Amongst the limitations of the study, there is no need to say, that further investigation is needed. The study was based on middle-class foreign students, excluding Argentinean peers. As such, there was no evidence of results in lower-classes or groups subject to durable psychological deprivations, such as students in zones of war or social conflict, as well as ongoing political instability or in devastated areas. PANCOE offers promising outcomes for the future of tourism marketing. Lastly, PANCOE shows an interesting potential to mitigate the negative effects of Covid-19 on consumers. PANCOE remains an innovative method that credits the important role of pleasure (joy) in tourism education.

After further review, the present case study brings reflection on the power of pleasure (joy) to better the classic education system. Whilst pleasure has been overlooked as an instrument to increase positive feelings in students for classic education, PANCOE (and the Laboratory of Joy) goes in the opposite direction. The Laboratory of Joy stimulates students' skills and performance through the articulation of pleasurable experiences. The results notably show that those students who participated in PANCOE obtained better degrees than those who did not taken part in the experiment. What seems to be more important, the endorphins liberated by positive interactive communication pave the way for the rise of pleasurable experiences, which, in turn, lead to better academic performances.

Last but not least, PANCOE was designed to standardise the learning process of foreign students. The main goal of PANCOE was to reduce the student dropout rate. The results showed that PANCOE not only improved the academic degrees of participants but facilitated their final graduation. Of course, more research should be focused on how PANCOE helps psychologically-disturbed students who live in unsafe destinations (war zones), conflict or disaster destinations, as well as whether the same results would be achieved for other groups. Although the outcomes are certainly innovative and exhilarating, they cannot be extrapolated to other universes simply because the method is not based on a statistical representative method. However, PANCOE was a successful experiment implemented by the Laboratory of Joy at the University of Palermo, Buenos Aires, Argentina.

Further reading

Airey, D., & Tribe, J. (Eds.). (2006). *An International Handbook of Tourism Education*. Abingdon: Routledge.

Amoah, V. A., & Baum, T. (1997). Tourism education: policy versus practice. *International Journal of Contemporary Hospitality Management*, 9(1), 5–12.

O'Keefe, T. (2014). *Epicureanism*. Abingdon: Routledge.

Sheldon, P. J., Fesenmaier, D. R., & Tribe, J. (2011). The tourism education futures initiative (TEFI): Activating change in tourism education. *Journal of Teaching in Travel & Tourism*, 11(1), 2–23.

Wang, J., Ayres, H., & Huyton, J. (2010). Is tourism education meeting the needs of the tourism industry? An Australian case study. *Journal of Hospitality and Tourism Education*, 22(1), 8–14.

Case 2

MARKETING WELLNESS IN PARADISE

What's hard about that?

Su Gibson and Drew Martin

Duration

One hour to one week, depending on course level and detail required for market strategy student presentations.

- Viewing the case video: 6.5 minutes
- Reading the case: 20 minutes
- Brainstorming and discussion: 30 minutes
- Conducting additional online research (optional)
- Working in groups, drafting of marketing strategic plan and presentation preparation (optional)
- Presenting individually or in groups (optional)

Learning objectives

Upon completing the case, participants will be able to:

- analyse dynamic customer behaviour using Maslow's needs theory
- develop a working knowledge of how customers' service quality assessment dimensions affect overall satisfaction for the lodging industry
- prioritise marketing goals when presented with challenges such as limited budgets, staffing shortages and changing guest expectations
- take a holistic view of marketing strategy and relate the example to operational concerns
- connect customer expectations and satisfaction with marketing efforts.

DOI: 10.4324/9781003182856-2

Target audience

A basic understanding of marketing and service management concepts is helpful for this case. However, the case study can be adapted for all course levels. Either teams or individuals can develop suggestions for marketing strategy and service quality initiatives.

Teaching methods and equipment

Resources for this case include a short video interview with the Resort's Vice President to inform the students and direct their work within a service management and marketing framework. A computer with internet access and audio/visual capabilities is required for viewing.

> Gibson, S. (2021, May 20). Interview with Jill Boyd of Hilton Head Health [Video]. YouTube https://youtu.be/zpWbiNUUSI4

Teaching methods could include small-group brainstorming of ideas for service improvement, in-class discussion about the importance of word-of-mouth advertising and online reputation management, peer-to-peer interviews about personal travel experiences, onsite visits to comparable resorts to observe changes in service offerings and guest behaviour, searches and discussions of online reviews with comparisons of pre- and post-pandemic reviews for the Resort and comparable resorts, and production of detailed marketing plans which may include activities aimed at improving customer satisfaction for the Resort along with justification for each initiative.

Teaching instructions

Suggested steps for the case are as follows:

1. Students watch the short interview video of Jill Boyd individually or as a class.
2. Students read the case.
3. Students search online, individually or in groups, for more information about the Resort, its branding, marketing messaging, and guest reviews.
4. Students read additional resources to inform their work.
5. Students present detailed marketing strategy to the class, based on the facts given in the case and their subsequent research.

Key concepts to explore before the case or while examining the case as a class include:

- Maslow's Hierarchy of Needs
- SERVQUAL

- gap analysis
- human resources, recruitment and staffing for specialised hospitality environments
- managing demand in a seasonal location.

Students could be asked to investigate the following:

- current occupancy levels of resorts in an area or nation
- trends in travel related to Covid-19 over time
- staffing levels, including nationwide unemployment rates, industry statistics, and local news related to hospitality staffing
- the ethics of promoting travel during pandemic conditions.

Possible student pitfalls include taking a short-term approach to strategic plans, underestimating the value of skilled employees and the negative consequences of employee turnover, overreliance on drawing new customers rather than retaining repeat guests when developing marketing strategy, and making the assumption that current environmental conditions remain static for the duration of their strategic plans.

Case

In the US, the travel and tourism industry lost an estimated US$766 billion in 2020 due to factors related to the Covid-19 pandemic (WTTC, 2021). Starting in March 2020, the lodging industry experienced a dramatic decline in customers. By mid-2021, leisure travel started to improve, but business travel continued to lag with average occupancy rates just above 50 percent. This occupancy level is an improvement from 25 percent the previous year (STR, 2021a, 2021b). As the lodging industry recovers from the global pandemic, marketing managers need to develop strategies under unfamiliar conditions. For a resort offering premium services, the cost of damaging customer goodwill is high.

Hilton Head Island, US in the South Carolina Lowcountry is known for its pristine white beaches, wooded biking trails, beautiful resorts, and exclusive golf and country clubs. A high-end destination for families and retirees, the Island has been home to Hilton Head Health for more than four decades. Guests typically stay for several weeks, and many are repeat guests. The resort is an all-inclusive weight-loss and wellness destination, offering guests custom-tailored wellness programmes, fitness classes, coaching, healthy living education, cooking demonstrations, and spa services. Jill Boyd, Senior VP of Hospitality for the Resort, is examining the marketing plans for the upcoming season. Due to financial pressure resulting from the Covid-19 pandemic of 2020, this year's marketing budget is reduced. Planned marketing activities must be especially effective to make use of limited resources. Several external environmental areas affect planning. The most pressing concerns are high customer demand and expectations.

Ms. Boyd currently notices trends in guests arriving with very high expectations, possibly inflated by extended periods of desire to travel to the Resort without being able to do so because of travel restrictions and time spent on the Resort's waitlists. Currently, these waitlists are necessary due to high demand and limitations in capacity imposed by the Resort to maintain a comfortable distance between guests.

Customers' expectations serve as performance standards with which guests measure their perceptions of service experiences. Perceptions of the service delivery are measured by customers' subjective assessments. The difference between customers' service expectations and perceptions is called the Customer Gap (see Parasuraman, Zeithaml, & Berry, 1985). Service providers can close the Customer Gap by understanding what customers want and how they assess the service offering. Understanding customers' expectations and perceptions help service providers design more effective services.

Maslow's (1943) hierarchy helps explain consumers' motivations to fulfill unmet needs. These needs are divided into lower and higher levels. Lower-level needs include physiological and safety needs. In a hospitality context, physiological needs are food and shelter. Promotional messages suggesting that a restaurant offers a filling meal, or a hotel provides only the basics and no more, aim to fulfill the target audience's physiological needs. Safety needs include security or avoidance of the unexpected. A hotel dedicating one floor to only women travellers addresses the safety concerns of a female guest who travels alone. Ms. Boyd notes that although pandemic-related restrictions have been lifted by local government, incoming guests are still wary of crowded facilities and have safety-related concerns.

Higher-level needs include social, esteem, and self-actualisation needs. To address social needs, hospitality services try to show that their customers are part of the community. A restaurant's promotion suggesting that you are treated like family adds a social element to the service delivery. Some of Hilton Head Health's success lies in strong bonds formed by guests to each other and to Resort staff during their stays. These bonds often continue after the guests leave the Resort. To address a customer's esteem needs, the hospitality provider must metaphorically roll out the red carpet. Customer databases maintained by hotels allow the service provider to make room adjustments based on prior visits and greet guests by name when they arrive. A customer's need for self-esteem reinforcement is addressed by access to premium benefits not available to most customers. Loyalty programmes offering upgrades provide a higher level of service for frequent customers. Luxury hospitality brands also address self-esteem needs. Finally, appeals to address self-actualisation needs offer a service experience allowing for the customer's personal growth. For example, Hilton Head Health's wellness education and programming aims to provide long-lasting self-improvement strategies to guests.

As people start traveling again, the hotel industry has an uphill battle to improve customer satisfaction. A recent report by the American Customer Satisfaction Index (ACSI) finds nearly a four-percent decline in customer satisfaction compared to pre-pandemic data (ACSI, 2021). While four percent does not seem significant,

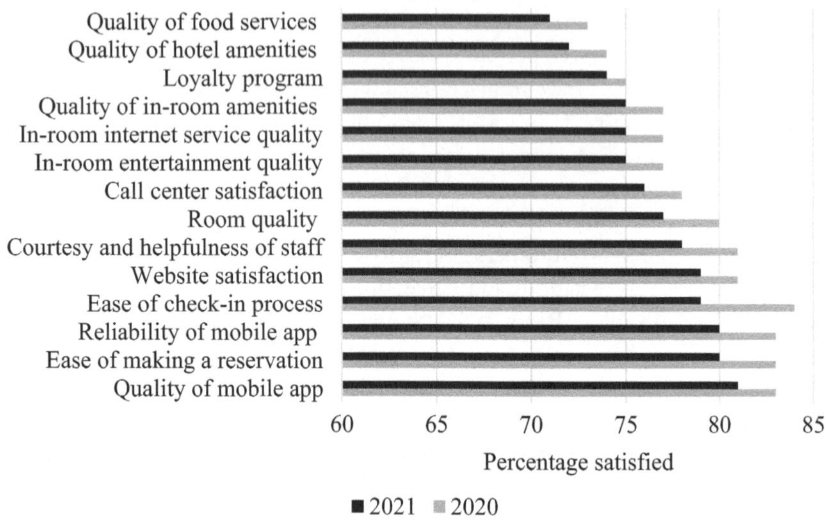

FIGURE 2.1 Customer Service Benchmarks for Hotels by Year

Source: ACSI, 2021

customer satisfaction with airlines has improved during this same timeframe and restaurants' customer satisfaction remains unchanged (ACSI, 2020, 2021). A closer look at customer satisfaction with hotel services shows declines in all areas. Jill Boyd sees similar results at Hilton Head Health, citing decreased facility capacity and lower staffing levels as possible sources of guest discontent.

Figure 2.1 shows customer satisfaction with services at hotels for 2020 and 2021. The maximum score is 100 and reflects the percentage of people surveyed who feel satisfied with a particular service element. For example, a score of 81 means 81 percent of people surveyed report satisfying experiences (19%, therefore, are unsatisfied).

One additional challenge facing Hilton Head Health is hiring more employees. Compared to 2020, South Carolina's leisure and hospitality industry has 30,000 fewer workers (a decline of 12.8%) (SCDOC, 2021). Optimal staffing has always been a challenge for the Resort, but the current worker shortage is severe. These decreases in staffing levels have slowed the Resort's recovery, limited guest capacity, and affected the level of perceived service quality experienced by guests. The current shortage is affected by workers changing careers, national legislation extending unemployment benefits that exceed some hospitality workers' base salaries, remote learning in schools creating childcare issues, and concerns about Covid-19 exposure. To recruit qualified workers, the industry is forced to offer higher wages and more flexibility than in the past. If more workers cannot be attracted, the industry needs to consider changing operational procedures.

Marketing messaging helps to shape customer expectations. Effective messages reduce the Customer Gap. Because Hilton Head Health has a high percentage of return guests, some additional communications may be necessary to inform those guests of recent changes to the customer experience.

A critical element influencing customer perceptions is guests' service quality assessments. Service quality assessments likely influence overall customer satisfaction as well. SERVQUAL identifies five dimensions of service quality (see Parasuraman, Zeithaml, & Berry, 1988). The *tangibility* dimension measures customer attitudes about the appearance of the facility, communication materials and personnel: 'Is the hotel's lobby attractive?' Measures of service delivery *reliability* are the second dimension: 'Was the hotel room available at the check-in time?' A third service quality dimension is *responsiveness* – providing prompt service: 'Was room service delivered in a reasonable time?' Assessment of how the service provider demonstrates competency, courtesy, and security measure the *assurance* dimension: 'Are security people visibly present at the hotel?' Finally, the *empathy* dimension is critical because service delivery is not always smooth: 'Is someone available to listen to the customer and understand their concerns?' As reducing the Customer Gap is critical, addressing service quality concerns are a step in the right direction.

Questions from Hilton Head Health

1. How have hotel customers' needs changed post-pandemic? Think of Hilton Head Health's guests in terms of Maslow's hierarchy. Provide evidence to support observed behavioural changes.
2. What are examples of best industry practices that the Resort should integrate into operations to improve service quality?
3. How can Jill Boyd improve guest satisfaction when designing the Resort's marketing strategy?
4. What other elements would be most critical to include in a marketing plan, given the current climate? Provide evidence to demonstrate why this area needs monitoring.
5. What are the ethical considerations of marketing to increase occupancy when pandemic conditions still exist in the local area?
6. What consequences might the pandemic recovery have on Hilton Head Health's brand?

References

American Customer Satisfaction Index (ACSI). (2020, October 27). American customer satisfaction index special COVID-19 restaurant study 2020. www.theacsi.org/news-and-resources/customer-satisfaction-reports/reports-2020/acsi-special-Covid-19-restaurant-study-2020/acsi-special-Covid-19-restaurant-study-2020-download

American Customer Satisfaction Index (ACSI). (2021, April 27). American customer satisfaction index travel report 2020–2021. www.theacsi.org/news-and-resources/customer-satisfaction-reports/reports-2021/acsi-travel-report-2020-2021

Maslow, A. H. (1943). A theory of human motivation. *Psychological Review*, 50(4), 370–396.

Parasuraman, A., Zeithaml, V. A., & Berry, L. L. (1985). A conceptual model of service quality and its implications for future research. *Journal of Marketing*, 49(4), 41–50. https://doi.org/10.1177/002224298504900403

Parasuraman, A., Zeithaml, V. A., & Berry, L.L. (1988). SERVQUAL: A multiple-item scale for measuring consumer perceptions of service quality. *Journal of Retailing*, 64(1), 12–40. https://doi.org/10.1177/002224299205600205

South Carolina Department of Commerce (SCDOC). (2021, April). *Economic Outlook*, 14(3). www.sccommerce.com/sites/default/files/2021-04/Economic_Outlook_April_2021.pdf

STR. (2021a, April 29). U.S. market recovery monitor – 24 April 2021. https://str.com/data-insights-blog/us-market-recovery-monitor-week-ending-24-april

STR. (2021b, May 4). Trend Report – United States. Retrieved from STR.

World Travel and Tourism Council (WTTC). (2021, March). United States: 2021 Annual Research. https://wttc.org/Research/Economic-Impact

Further reading

Breier, M., Kallmuenzer, A., Clauss, T., Gast, J., Kraus, S., & Tiberius, V. (2021). The role of business model innovation in the hospitality industry during the COVID-19 crisis. *International Journal of Hospitality Management*, 92, 102723. https://doi.org/10.1016/j.ijhm.2020.102723

Chan, J., Gao, Y. L., & McGinley, S. (2021). Updates in service standards in hotels: How COVID-19 changed operations. *International Journal of Contemporary Hospitality Management*. http://dx.doi.org/10.1108/IJCHM-09-2020-1013

Gibson, S. (2021, May 20). Interview with Jill Boyd of Hilton Head Health [Video]. YouTube https://youtu.be/zpWbiNUUSI4

Knutson, B., Stevens, P., Wullaert, C., Patton, M., & Yokoyama, F. (1990). LODGSERV: A service quality index for the lodging industry. *Hospitality Research Journal*, 14(2), 277–284. https://doi.org/10.1177/109634809001400230

Case 3

MENTOR-MENTEE INCENTIVE EVENT PLANNING

Özen Kırant Yozcu and Maria D. Alvarez

Duration

One semester (approximately 12 weeks) in parallel with theoretical classes that will serve as a foundation.

The groups are expected to meet with their mentors every week for a minimum of one hour discussions. The first part of the project (creation of an event/meeting company) is designed to be completed within three weeks. The brief for the actual event to be developed is provided on the fourth week by the partner (in our particular case, MPI Turkey). Following this, the mind map of the event project is prepared in seven weeks. The final week is assigned for the group presentations, during which the student groups need to present their event in 15–20 minute slots.

Learning objectives

Upon completing the case, participants will be able to:

- better understand the intricacies of event planning through hands on experience
- obtain a view of event management as an enjoyable activity and the events/meeting industry
- become more aware about sustainability and corporate social responsibility activities (CSR) within the framework of event planning
- become part of a network that includes mentors from the industry, which is expected to facilitate the employment of students and staffing of key industry positions, in a win-win situation.

DOI: 10.4324/9781003182856-3

Target audience

The target audience for the case is mainly senior undergraduate students that are pursuing a bachelor's degree in tourism management or related fields. The case requires participants to be familiar with the workings of destinations and of the tourism industry in general. It also assumes that students will have some knowledge of strategic marketing principles. For this reason, it is recommended that the case is used to teach students that are in the final year of their bachelor's degree. In addition, the case also aims to integrate students with a network of event/meeting industry professionals. Therefore, it is suitable for students that are near their graduation date, who may best benefit from such a network. The case was applied online to senior students of the 'Conventions and Special Events' course in the fall of the 2020–2021 academic year.

The case is suitable for group work and is designed for 25–45 participants that will be divided into groups of three to four students. Each group is paired with a specific mentor, a constraint that may limit the number of participants.

Teaching methods and equipment

The case can be implemented online, offline via in-classroom education, or in a hybrid format. The case requires the cooperation of professionals in the field of event marketing, who can act as mentors for the student groups.

The groups meet with their mentors on a weekly basis to discuss each step of the event project and to brainstorm within the group. At each step, the group is required to report to the instructor using the mind mapping technique, which is used for gathering and ordering information, as well as for problem solving, since it helps form more creative ideas and find solutions (Buzan, 2006). The students start by putting the event case topic in the centre of the box in the mind map and continue to lay out the details given in the request proposal form, such as hotel preferences, transportation details, food and beverage activities, etc., using the branches emanating from the central box. This helps students to see every detail of the event case, while also looking at the big picture.

The case concludes with the group presentations of the students in front of a jury consisting of all the mentors. Different presentation techniques can be used by the students to present their findings in a professional manner and students are encouraged to be creative. The performance of the best group or groups can be rewarded with professionally-related pre-specified prizes that are provided by the mentors.

If the case is to be implemented online, all meetings of the students among themselves, those with the instructor and mentors, as well as the final presentations, can be carried out online via available online meeting platforms.

This case was implemented by the first author during the 2020–2021 fall semester in cooperation with an industry partner. In this situation, MPI (Meeting Professionals International) Turkey Chapter was the partner for this mentor-mentee incentive event planning case study. The members of the highest ranked group were rewarded with a complimentary membership to the MPI Turkey Chapter.

Teaching instructions

The case consists of four main steps. The first step requires the creation by the student groups of an event/meeting company. The second step is related to obtaining the brief from the partner company (in our case, MPI Turkey Chapter). The third step requires the students to prepare a mind map of the event and the fourth step is about finalising and presenting the final proposal for the specific event.

Step 1

Students organise their groups and a mentor is assigned to each group at the beginning of the course. The interaction of the students with members of the industry provides an opportunity to add to the theoretical lectures that are provided by the instructor. The mentors meet with the students on a weekly basis to guide them through the process. In particular, through the case, students identify the importance of combining analytical methods, such as benchmarking, stakeholder analysis, SWOT analysis, etc., with practical considerations relating to the event/meeting industry.

So, as a first step, students create their event company defining their vision, mission, goals and objectives. They caryy out a SWOT analysis to understand the company's situation; a benchmarking analysis to look at best practices from the competitors in the market; a stakeholder analysis to define internal and external partners; and a PESTELI analysis to explore the details of the political, economic, social, technological, environmental, legal and industry situation. The groups also try to create scenarios to analyse future potential situations. As part of the stakeholder analysis, students research who their partners in this event case are (the organisations with whom they will work for this case).

For example, students need to decide about the hotel where the event will take place, the transportation vehicles to be used, the entertainment company that will provide the entertainment part of the event, the company that will offer engineering and technology solutions before and during the event, the decoration partners for the gala dinner of the event, etc. So the stakeholder analysis should be conveyed to the students as critical for the success of the event. In addition, different scenarios should be taken into consideration while planning the programme and identifying the stakeholders. The students study worst case scenarios, including risk analysis, such as potential natural risks like bad weather, diseases like the Covid-19 pandemic, or any other situations that may lead to cancellations and technological problems.

Step 2

The second part of the event project is related to receiving and analysing an extended brief from the project partner. A meeting is organised for the members of the partner organisation to explain the details of the event project and their expectations from the groups. The students are asked to prepare the event project as if they would be bidding for the event of XXX Company.

Step 3

Every week after the instructor provides theoretical lectures on the specific topics in class, the students discuss these issues and their implications for the industry and their particular event with their mentors. As a reference point for the students, the relevant chapters from Fenich (2015) are provided. These include strategic planning, events as complex projects, identification of relevant stakeholders and of the nature of relationships with them, creation of a risk management plan and an implementation plan. The usage of technology is another topic that needs to be included by the students and therefore it also needs to be added to the theoretical lectures given by the course instructor. Furthermore, legal issues, including contracts with the customer and the stakeholders, are studied.

As taught to the students, planning of an event programme starts with creating a concept of the event, and continues with the preparation of the timeline and agenda. As such, during the event planning stage, students evaluate proposals from their stakeholders to find out the best one for the available budget. In addition, they learn how to link the theme of the event to their operations. In particular, the students understand the importance of target market analysis to find out the wants and needs of the guests in the event project so they can use the most suitable option. In the event case, food and beverage related decisions are very important as the guests will come from another country, and will generally require unique menus customised for them. Therefore, different restaurants and venues need to be proposed by the students.

If the case is related to a corporate event, as in the event for XXX Company, social responsibility activities (CSR), public relations and digital marketing tools need to be included to market the event. This may include specific activities carried out during the meeting, like creating a memorial forest on behalf of XXX Company, organising charity actions, or focusing on sustainability issues throughout the event programme, such as choosing green certified hotels, avoiding the use of plastic materials, controlling waste management, preferring reusable and recyclable materials and gifts. These CSR activities may also be planned as something that may be shared by the guests' social media accounts or used by public relations for brand awareness.

Furthermore, the students learn to integrate different marketing communications tools within the event planning exercise. For example, to create brand awareness, it could be suggested that products of XXX Company be included in the rooms of the guests as hotel amenities or specially designed gifts may be delivered before the arrivals of the guests. Also, the students may explore the use of influencers in their activities. The use of social media in achieving the objectives of the events should also be considered by the groups, as they may plan for different ways of integrating social media into the event. For instance, the plan may include a way to encourage guests to share sightseeing activities in their social media accounts.

All of these issues that are covered in the theoretical part of the course also need to be taken into consideration by the students when preparing their event plan. They are expected to assign all these details into their mind maps. In addition, after a four-week interval, the groups are required to meet with their instructor to explain the steps and their progress.

Step 4

In the final part of the event project, groups should have finalised all the required steps and prepared a proposal form that includes issues such as hotels, transportation, entertainment and budget. These details are presented to a jury of mentors from the partner organisation.

The results of mentorship programmes can have positive outcomes related to career, attitude, motivation, behaviour and relationships (Hamilton, Boman, Rubin & Sahota, 2019). This case from the event industry gives students an alternative and comprehensive view of the organisation of an event. The participants are able to understand the processes more clearly and realise the difficulties involved in organising events/meetings. The case is also thought to convey to the students that event management is a challenging and enjoyable part of the tourism industry.

At the end of the semester, feedback on the implementation of this case was obtained from the students. Many commented on the learning experience and the excellent relationship that they were able to build with their mentors. The mentors' advice was described as being very valuable to understand the practical implications beyond what is covered by the theory in the course. Moreover, some students claimed that the course and the case helped them to change their opinion about the event industry and encouraged them to seek employment in this industry. Several of the students remarked that networking with the members of MPI Turkey allowed them to easily obtain job interviews for positions after graduation.

Case

The case was obtained by an important and experienced meeting company in Istanbul. The company is a professional reputed organisation that welcomes incentive groups from different countries. The case is based on one of their prior projects and is shared here with their consent. The details of the case are given below.

The event company is participating in a competition for an Incentive Group event of XXX Company (working in the fast moving consumer goods industry) in YYY Country. It is one of the most important Incentive Group events for XXX Company, and as such has been held annually for 15 years. The participants invited are their main customers – owners of large supermarkets, drugstores and distributors in YYY Country. The guest profile is a class 'A' group, with high purchasing power, used to traveling and staying in luxury hotels. Therefore, the XXX

Company proposal is not only to take the guests on a luxury trip but to make this trip a time of networking and recognition, while providing a learning experience and knowledge. The theme for this event will be *Negotiation and Differentiation*.

The trip duration will be six nights / five full-days at the destination. Flights have not been defined yet, but arrival may be in the morning or in the afternoon. The total number of participants is 90 people (60 guests and 20 members of the senior management of XXX Company in YYY Country – ten of them will be participating with their spouses or significant others). In terms of their characteristics, 80% are men aged between 40 and 60 years of age, modern and demanding executives. The date is not set but the suggestion is that the event would take place during the first or second week of September. A suggested programme, presented with the relevant costs is requested. Whether costs are net or gross, include taxes, service charges and gratuities needs to be specified.

Programme requirements are as follows:

- *Hotels*: At least three options of very luxurious and modern hotels – the suggestion is to provide new hotels, with a high level of luxury and technology – should be offered. In previous years they have stayed at the Royal Monceau in Paris and Corinthia in London. 60 single rooms (guests) + ten single rooms and ten double rooms (XXX Company Directors with their spouses or partners) are needed. Details concerning early check-in and late check-out policy, porterage and gift delivery fees need to be provided.
- *Transfers and Transportation*: For local transportation, luxury and private coaches with mineral water and YYY language bilingual guides (if possible) at all times need to be considered. On arrival and departure at the airport, the luggage should be transported separately from the group and should arrive at the hotel before the participants.
- *Meeting Requirements*: There will be one full-day or two half-day meetings / workshops for the group.

  Suggestions for keynote speakers for one hour in the morning and another one during the meeting should also be included. The meeting could take place at the hotel, at a university or at some other site that can be unique to the group (different suggestions and alternatives need to be provided). The group is YYY-speaking. The proposal should take advantage of local skills, such as experience in negotiation, and invite an important trader/entrepreneur to give a lecture or a testimony about his/her experience, explaining about negotiation strategies during the workshop or during an external activity.
- *Sightseeing programmes and social activities*: Alternatives for a social programme, including some local activities, cultural and historical tours, need to be suggested. These should include some experiences which the guests cannot find/book by themselves. The sightseeing programme may also involve a trade visit – provided by XXX Company. In any case, a transfer service should be arranged for participants. When determining the details and contents of a

sightseeing programme, a fit between the event theme and social activities/trade visits should be sought.

- *Lunches and dinners:* A welcome reception and a gala dinner should be included. The gala dinner will be the farewell dinner and should be very unique. In previous groups, Chateau de Versailles in Paris and a castle in London were used privately by the group. A very special, unusual and private venue – thinking about music, decoration, cocktails, dinner, etc. – should be suggested. Other meals can be local or international cuisines, or a combination of both.

Also proposals for typical and very special gifts should be provided. Other required suggestions include bars or night clubs after dinner and cultural experiences. In addition, tour options for the spouses and partners of the XXX Company's directors are expected. Within this programme, CSR activities should be organised.

A site inspection visit that includes three people (client and agency staff) should be planned. In addition to the best date for the visit, an estimated cost should also be provided.

Contingency plans should be developed for any risks before, during, and after the event, including plans related to first aid, crowd management, other security and safety issues, and technology-related problems. Also, the situation where the group or some of the participants cannot travel due to the Covid-19 pandemic should be considered, so a plan for this situation and suggestions for alternative digital content need to be included.

References and further reading

Arnesson, K., & Albinsson, G. (2017). Mentorship – a pedagogical method for integration of theory and practice in higher education. *Nordic Journal of Studies in Educational Policy*, 3(3), 202–217.

Buzan, T. (2006). *Mind Mapping: Kickstart Your Creativity and Transform Your Life*. England: Pearson Prentice Hall.

Fenich, G. G. (2015). *Planning and Management of Meetings, Exposition Events and Conventions*. England: Pearson Prentice Hall.

Hamilton, L. K., Boman, J., Rubin, H., & Sahota, B. K. (2019). Examining the impact of a university mentorship program on student outcomes. *International Journal of Mentoring and Coaching in Education*, 8(1), 19–36.

Case 4

MARKETING YOUR DESTINATION FROM A THOUSAND MILES AWAY

The power of social media

Elaine Yulan Zhang and Carol Yi Cui

Duration

120 minutes or more.

Learning objectives:

Upon completing the case, participants will be able to:

- evaluate digital marketing performance on social media
- conduct competitive analysis for tourism products in digital marketing
- formulate a marketing plan and its presentation as marketing professionals
- formulate communication strategy for tourism products on social media.

Target audience

This case aims to introduce competitive analysis as a tool in developing a marketing plan and the principles of successful social media strategies. The knowledge and skills are applicable to organisations in tourism that wish to use social media as a tool to develop their customer base. This case study can be used in higher education and vocational programmes. It will be an interesting and effective exercise during workshops after a lecture on the importance of social media strategies. After completing the case, students will understand better the process of developing an effective marketing plan.

The case is designed for a group of 18–24 participants and requires participants to be divided into six teams. The instructor groups participants into teams of three to four members and assigns one of the four roles to each team: (1) DMO for Vinicunca – Commission for Promotion of Export and Tourism (PROMPERU);

DOI: 10.4324/9781003182856-4

(2) DMO for Zhangye National Geopark – Gansu Tourism Bureau; (3) digital marketing agencies (at least two teams representing two agencies) who provide a marketing plan to promote Vinicunca; and (4) digital marketing agencies (at least two teams representing two agencies) who provide a marketing plan to promote Zhangye National Geopark. If the class size is larger, more teams representing agencies can be added. If the class size is smaller, one attraction can be selected, and three teams can be formed: one DMO team and two agency teams. Competitive analysis can still be carried out, considering the other attraction as a potential competitor.

Teaching methods and equipment

This case uses role play, competitive analysis and group discussion as the teaching methods. The scenario is that marketing agents present their ideas for marketing plans to DMOs, who will select their preferred agent. To effectively execute the activities, a classroom can be used for students to conduct the analysis and group discussion, and later, a meeting room setting can be arranged for the presentation. Students are required to carry out analysis with not only the information provided in the case but also the most recent social media marketing activities they have observed. Thus, students are expected to have devices (e.g. computer, smartphone) to access the internet to complete the tasks. If the classroom is equipped with computers and projectors, the marketing agency teams are encouraged to use them during the presentation.

Teaching instructions

Terminology

This case aims to establish practical knowledge of terms for tourism students when facing social media marketing challenges. Here are key terms to be communicated through this case:

Social media strategy: A social media strategy is a summary of everything you plan to do and hope to achieve on social media. A social media strategy is usually presented as a pitch book format in slides, and should describe the 4Ps of the social media plan and define a specific marketing goal and how it will be achieved.

KPI (key performance indicator): This is a numeric description of your marketing goal, e.g. 'This year's social media KPI is to achieve 1 million fans on Facebook.' A KPI should not just be the number of fans, it can also be about video plays, content variety, impressions and search engine rankings.

Target market: Finding a target market is an important job when sorting out marketing objectives for a tourism destination. A market should be described geographically, demographically and socio-culturally, with an estimation of market size. It can also be further developed into a consumer persona by providing a semi-fictional depiction of your ideal buyer based on market research and actual data about your present consumers.

CPA (cost-per-action): CPA can also be interpreted as cost-per-acquisition or cost-per-download. It is a useful metric to measure your marketing campaign's effect, by using all the cost invested for the campaign deducted by all the actions you gained from it. Actions here could be download / like / participation / purchase / writing reviews / shooting videos or any action fit. CPC is similar to CPA, but it only refers to cost-per-click.

CPM (cost-per-thousand-impressions): Similar to CPA, CPM is a useful metric to measure your marketing campaign's awareness effect. The benefit of using CPM is that it fits all media as a marketing campaign's ultimate goal is to be seen. The commonly used minimum unit for online awareness is 1000 impressions.

Organising teaching

The following three steps are explained with the case being executed with a group of 18 participants. The steps can be adapted to different class sizes and the time needed for preparation and presentation can be altered according to the background of students and the number of marketing agency teams. Before the execution of the case, instructors can introduce the importance of having clearly defined social media strategies together with a marking plan, and also the importance of learning-by-doing in digital marketing. Social media is the focus of this case and digital marketing can involve other existing and emerging digital platforms as well.

Step 1

Instruction and organising students are expected to take 5 minutes. Participants are randomly divided into six teams and each team has three members. Instructors can display the pictures of Vinicunca in Peru and Zhangye National Geopark in China (Figure 4.1) to show the very similar landscape and encourage students to search for information about these two attractions. The case material can be used here.

Next, instructors explain tasks for each team. All participants are expected to compare the content posted by both DMOs' social media platforms (Facebook, Instagram, etc.), and evaluate the performance of the existing digital marketing. The case material 'social media strategies' and 'How to use social media help attractions attract tourists' can be used here.

The four teams who represent agencies are expected to prepare a digital marketing plan for DMOs and a 5–10 minute presentation. The two teams who represent DMOs are expected to brainstorm criteria for choosing a digital marketing agent if they would like to outsource digital marketing activity. They need to listen to all the presentations, ask questions after presentations (5 minutes per agent), and present their criteria and final choice in 5 minutes. The case material 'Basic steps to formulate social media strategies for DMO' can be used here.

FIGURE 4.1 Zhangye National Geopark in China

All participants are given time (around 40 minutes) to complete the research and preparation for the presentation. Instructors may wish to give students more research time after class and allow them to present in the next class. Thus, this case study can be completed within a 2-hour in-class seminar if students are experienced in preparing a marketing plan, or be expanded to two to three weeks if students need more research and preparation time.

Instructors may highlight the following principles and guidelines before and during the students' discussion:

- When preparing the digital plans, students should pay attention to three principles (details are explained in the case):
  1. Value 'social' more than 'media': use social media to influence your target market.
  2. Smart strategy: use social media for niche marketing.
  3. Marketing principles work: social media is a tool in achieving marketing objectives.
- The content of the presentation should include:
  1. objectives of digital marketing
  2. current performance review of DMO on social media
  3. target audience
  4. proposed marketing plan with 4Ps
  5. sample social media contents
  6. budgeting.
- When choosing the right target market, students may consider three questions:
  1. What is the current market? Who is visiting right now?
  2. What is the potential market? What is the size of the potential market?
  3. Can you describe the potential market consumer persona by answering the below questions?
     a. What is their geographical information?

  b. What are their demographic stats?
  c. What are their social-cultural preferences?

- When preparing the budgeting, students may need to consider the average cost of advertising (e.g. CPC) and present the calculation (examples are included in the case) where the budget matches the social media objectives (e.g. number of impressions). Here is one example:

  From budget to objectives: For example, if you would like to increase the awareness of the attraction among an audience with a budget of US$10,000, you may consider using the CPM model, so the total coverage would be 10,000/7.19 (from Table 4.1) = 1,390,000 impressions.

- When preparing sample content for social media marketing, try to put yourself in the consumer's shoes:
  1. Prepare a seven-day content plan for your destination – its attraction, hotels, and transport must be mentioned.
  2. Prepare at least one festive campaign for your target market.
  3. Propose at least one key opinion leader to work with for your destination.

Step 2

In the second step, the two marketing agency teams for Vinicunca in Peru will present their marketing plans. The team representing the DMO of Vinicunca will ask questions and make a decision on the preferred marketing plan. The two marketing agency teams for Zhangye will present their marketing plans and DMO will ask questions and make a decision. The whole process will take around 60 minutes. While a team is presenting, other participants are encouraged to take notes on any attractive marketing ideas and effective social media strategies. All participants can put down their preferences after listening to the presentations and later compare their own choices with the DMO's final decision.

Instructors may highlight the following criteria DMOs may consider before making the final decision:

- whether the plan is complete
- whether the plan is consistent and self-justifiable
- whether the market sizing and budget are reasonable and practical
- whether the proposed sample content is liked by audiences present.

Step 3

The final step will be a classroom discussion guided by the instructor and that should take around 15 minutes After all presentations are delivered by DMOs and agencies, the instructor concludes the section by highlighting the expected benefits of using social media for remote attractions and innovative marketing ideas. All participants are encouraged to reflect on whether they have applied the rules and steps introduced in the case. Instructors will make comments based on their performance in step 2. After the class, participants are encouraged to share their ideas with DMOs via social media or other channels.

Instructors may highlight the following key issues during the guided discussion:

- Both the organisation itself and the marketing agency will need to understand the current practices on social media platforms in order to develop the new social media strategies.
- Social media is unique in terms of its influence on targets' social circle. Targeting the right person is more important than approaching a larger population.
- The social media strategies should match the preferences of targets and what the organisations can offer.

Case

Now, please open your favourite search engine and search for pictures of 'Vinicunca' or 'Rainbow Mountain' in Peru and 'Zhangye National Geopark' or 'Zhangye Danxia' in Gansu Province, China. You may be surprised by how the landscapes of these two attractions in two different countries are similar to each other. Both mountains are 'painted' with coloured layers. Not only in look, these two attractions are also similar in the following aspects:

- *Geography*: Both of them are located in high altitude areas. Vinicunca is a mountain in the Andes of Peru with an altitude of 5,200 metres (17,100 ft) above sea level. Zhangye is located in the northern foothills of the Qilian Mountains with an altitude of 2000–3800 meters (6,560–1,2470 ft).
- *Accessibility*: Accessibility to both attractions is limited. To visit Vinicunca, tourists need to take a two-hour drive from Cusco and a walk of about five kilometers (3.1 mi), or a three-and-a-half-hour drive through Pitumarca and a half-kilometre (0.31 mi) steep walk (1–1.5 hours) to the hill. As of 2019, no robust methods of transportation to Vinicunca have been developed to accommodate tourists, as it requires passage through a valley. To visit Zhangye National Geopark, it takes about one hour drive from Zhangye City. Flights to Zhangye are limited to a few domestic cities in China such as Chongqing, Lanzhou.
- *New to market*: Probably due to the limited accessibility, both attractions are relatively new. Vinicunca started to attract mass tourists in the middle of the 2010s. 'Zhangye National Geopark' was formally designated by the National Ministry of Land and Resources on 16th June 2016, after it has passed the on-site acceptance test. In 2019, Vinicunca received more than 1,500 visitors every day; and Zhangye, which covers an area of 322 square kilometres, received 7,800 daily visitors on average.

Social media strategies

Both attractions have been promoted by DMOs via social media platforms such as Facebook and Instagram. With a similar landscape, different marketing approaches can be observed from the content of information disseminated on the social media accounts of these two attractions. A comparison of the digital marketing performance of the two attractions can be conducted by looking at the strategies used and performance indicators. Their social media pages can be found below:

- Commission for Promotion of Export and Tourism at Visit Peru www.facebook.com/visitperu
- Instagram Fan Page www.instagram.com/visitperuofficial/
- Gansu Tourism Bureau at Gansu China www.facebook.com/GansuChina
- Instagram Fan Page www.instagram.com/gansu_china/

According to Munar (2012), there are five generic social media strategies used by DMOs: *mimetic*, *advertising*, *analytic*, *immersion* and *gamification*. A *mimetic* strategy is to make use of the style and culture in social media platforms to encourage sharing from tourists. It is challenging to make tourists feel real involvement in the community. Many Facebook posts at Visit Peru are featured with photos taken and shared by tourists on either Facebook or other social media platforms. This practice may encourage other tourists to share content during or after their visit. *Advertising* strategy is to gather the users on a social media platform for promotion purposes such as distributing news and advertisements. The majority of posts about Zhangye at Visit Gansu include pictures uploaded by DMOs, a typical practice of using social media as an extra free advertising board. *Analytic* strategy is to understand the image of a destination and the impacts of their social media activities by analysing the content and data available on social media platforms. It is difficult to observe whether DMOs are conducting analysis but this strategy can be carried out by a digital marketing agency who can potentially be hired by DMOs. *Immersion* is a strategy to develop a new community on a social media platform, such as Visit Peru and Visit Gansu on Facebook. These two DMOs have also established communities on other channels, for example, Visit Peru on Flickr (www.flickr.com/groups/visitperu), Peru Travel on Sina Weibo in Chinese (https://weibo.com/PeruTravel). *Gamification* strategy applies gaming technology to non-gaming sites to reach a wide audience. As of 11th March 2021, when this case was prepared, this strategy was not adopted by either Visit Peru nor Visit Gansu.

Performance of social media marketing is often measured by awareness, engagement and share of voice. Awareness of an attraction or destination can be measured by the impression, which tells how many times a post has been displayed to users on a platform. Often, the awareness of the post can be influenced by the number of followers or the size of the community on a social media platform. The engagement can be measured by indicators such as the number of likes received by a post, number of times the post is shared, number of comments under posts, number of times that other users @ or tag the DMOs, conversions between DMOs and other users. Share of voice can be understood by looking into what keywords are associated with this destination in discussion and the sentiments associated with the discussion. Most of these indicators can be accessed by external users, thus, can be used as marketing agents for analysis purposes. The two attractions are included in the content of both Facebook pages, however, the awareness and engagement of Visit Gansu are limited, thus the opportunities of understanding share of voice is also limited. The effectiveness of the two DMOs using each of the above strategies can be analysed.

How to use social media to help attractions attract tourists

Rule 1: Value 'Social' more than 'media'

Does social media attract tourists? Yes, but influence from DMOs is sometimes limited. To improve the performance of social media marketing, DMOs need to understand that the unique power of social media is not from 'media', but from 'social'. Digital marketing agents could help DMOs by providing clearly identified objectives, strategies and plans. To prepare a successful proposal of a social media marketing plan, fundamental marketing principles, such as considering the target audience, are still important. First, goals and objectives should be identified. After analysing the performance of both DMOs, major areas to improve can be identified. Second, the target audience should be clarified. Because of the location, the major target markets of Vinicunca and Zhangye may be different. Third, potential competitors should be explored. Attractions' competitors are not limited to attractions with similar landscapes. Fourth, the social media strategy should be proposed. The strategy should match the goals and objectives of the DMOs. Key performance indicators should be included to demonstrate how an agent will evaluate its performance.

Another similarity between the two attractions is the concern of environmental impacts of tourists on these natural attractions, for example degradation because of all tourists using the same trail path. Gansu government has been controlling the number of visitors according to an instantaneous carrying capacity of 5000, especially during peak seasons. How to utilise digital marketing tools, not only attracting more tourists but also nudging the tourists on smart plans and responsible travel, is another area to explore.

Rule 2: Select your strategy smartly

The direct impact brought by social media on destination marketing is the 'long-tail economy'. The long-tail model (Figure 4.2) is a business strategy that allows businesses to realise significant profits by selling low volumes of hard-to-find items to many customers, instead of profiting from economies of scale. The long-tail is a

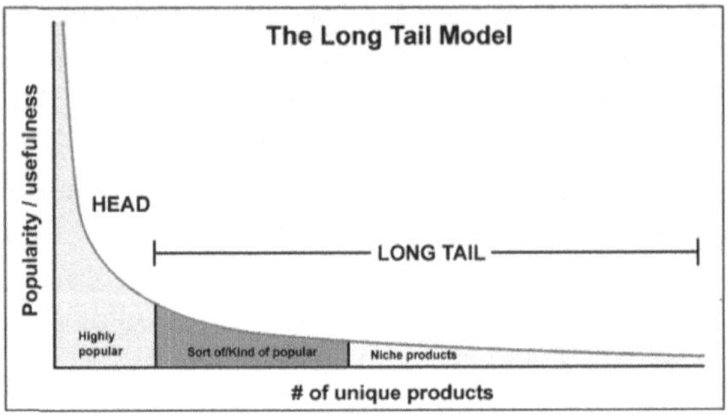

FIGURE 4.2 Long Tail Model

TABLE 4.1 Average Facebook Advertising Cost

Average Facebook Advertising Cost	*Bidding Model*
$0.97	Cost-per-click (CPC)
$7.19	Cost-per-thousand-impressions (CPM)
$1.07	Cost-per-like (CPL)
$5.47	Cost-per-download (CPA)

Source: WebFX (2020)

concept first raised by writer Chris Anderson in *Wired Magazine* in 2004. Internet and social media have enabled the long tail model to become an effective business strategy available to all. Consumers are also shifting from mass-market buying to more niche or artisan shopping. For tourism destinations, a long-tail model will bring opportunities to those remote destinations. By shifting strategic objectives, remote destinations could focus on the adventure lover market. Destinations like Vinicunca and Zhangye both fall into this category.

Basic steps to formulate social media strategies for DMO

Step1: Set up an objective by knowing CPM and CPC

Social media objectives include the *number of fans*, *number of interactions/deals* and *number of impressions*, as mentioned before. Social media uses a dynamic bidding model, meaning a business should find its optimal competition market within an affordable price range. Table 4.1 provides a summary of average ad cost on Facebook by AdEspresso (WebFX, 2020). By using this data, you can determine the total coverage achieved with your given budget of a campaign. With the total coverage known, you can try to locate the target market geographically and demographically. Can you try to formulate a market brief for Vinicunca and Zhangye, respectively, with a 100,000 USD budget on social media?

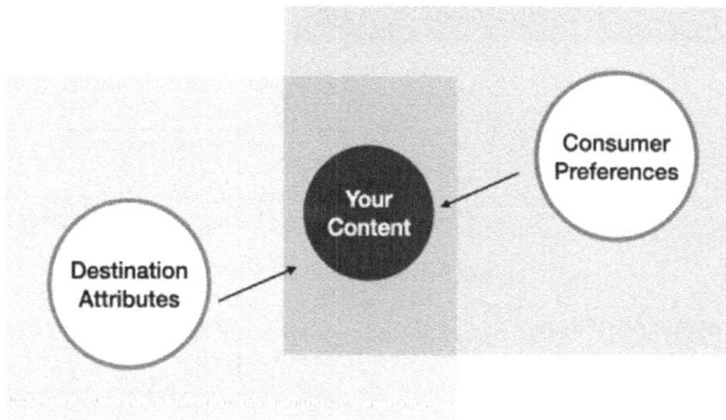

FIGURE 4.3 Attributes and Preferences Matching

Step 2: Formulate a content plate of your destination

Content formulation is a process to review your destination attributes, matching them with target consumer preferences (Figure 4.3). Therefore, different target markets require customised content design. You can easily find some social media content ideas from search engines. But most important is to match popular ideas with your destination. Go through the social media users' preferences below and try to formulate a table for Vinicunca and Zhangye respectively. For example, those who purchase from high-end sports apparel brands may be worth targeting since they may enjoy more nature-based activities like hiking.

Social media users' preferences:

- apparel
- auto
- beauty and makeup
- fitness
- sports
- home improvement
- food and beverage
- luxury products
- nature and photography
- art and culture
- festivals and special dates.

Step 3: Motivate your fans on social media

As mentioned before, 'social' is more efficient than 'media' use of the internet. 'Social' is the key to successfully leverage the power of digital marketing. Every successful social campaign should motivate fans. There is no need to be intimidated by high-value prizes like a 'million-dollar job'. Any travel souvenirs of the destination could be a perfect motivation elsewhere.

	Strategy Formulation		Marketing 4P Implementation Plan
DMO	Objectives		Price
	Target Market	→	Product
	Consumer Preferences		Place
Set a Budget	Destination Attributes		Promotion

FIGURE 4.4 Strategy Formulation and Implementation Plan

It is reasonable to wonder, where shall influencers fit in this strategy? Celebrity should be seen as an intangible motivational power to your campaign. The idea of using an influencer is to convince his/her fans to join your campaign. Try to convince an influencer to become your fan and promote you. Be the host of your own campaign first.

Present your DMO strategy to the class

Copy and complete the strategy formulation and implemation plan in Figure 4.4 and present your DMO strategy to the class.

References

Munar, A. M. (2012). Social media strategies and destination management. *Scandinavian Journal of Hospitality and Tourism*, 12(2), 101–120.

WebFX (2020). How Much Does Facebook Advertising Cost in 2021? www.webfx.com/social-media/how-much-does-facebook-advertising-cost.html.

Further reading

Pike, S. (2015). *Destination marketing: Essentials*. New York: Routledge.

Case 5

USING ONLINE REVIEWS TO MEASURE CUSTOMER EXPERIENCE ASSESSMENT

Andrés Artal-Tur, Juan Pedro Mellinas and Noelia Sánchez-Casado

Duration

The case study takes place over three different sessions – one in the classroom and two in the lab – of around two hours each. The first session will present the structure of the study as well as the concepts to be learned. The second session will illustrate the use of online reviews from tourism platforms to examine the marketing topics under study. This includes the relationship between theory and empirics, as well as techniques for downloading and organising the dataset in the case study. The third session will focus on analysing the data by means of statistical methods, followed by a classroom discussion on the main findings and conclusions.

Learning objectives

Upon completing the case, participants will be able to:

- identify and understand the theoretical marketing framework on customer experience value assessment that will determine the structure of the case study
- learn how to select and download data from tourism platforms
- prepare the dataset for further analysis, while relating it to the theoretical concepts previously reviewed
- develop the statistical analysis of the data using the corresponding software, while running the defined tests
- compare the results obtained by different class groups and discuss the main findings and conclusions emerging from the analysis.

DOI: 10.4324/9781003182856-5

Target audience

The case study is for students in the first year of business, economics, tourism, or marketing degrees. However, it could also be of interest for postgraduates. To follow this case study, some background knowledge of two topics is required: (1) introductory statistics, including descriptive statistics and basic tests, such as the analysis of variance (ANOVA); and (2) social marketing, including online review assessment and customer experience value theory.

Teaching methods and equipment

The case study is well-suited for in-class modelling design, as well as for group work. The teaching methodology provides initial information gathered from a tourism platform, which compares the assessment of two key issues from visitors to the two leading theme parks in Spain. After being provided with this set of information, the students are required to design and run an analytical method to obtain robust conclusions on customer perception and online related posts. The analysis requires the use of introductory statistical methods using standard computer software.

Teaching instructions

Step 1 (Session 1)

Define a theoretical framework and discuss with the students the main concepts to be learned.

Step 2 (Session 2)

Present the data to be found, download it from the tourism platform and organise the dataset according to the theoretical concepts introduced in Step 1. Prepare the data for further statistical analysis. The data used for analysis during the preparation of this case can be accessed on the following website in the forms of Excel sheets: https://upct-my.sharepoint.com/:x:/g/personal/andres_artal_upct_es/EUM1QE--amxBpsoRrhOMqVEB42zSDFEAytokBurfb006ww

Step 3 (Session 3a)

Run the statistical ANOVA test on the selected software, explain the procedure and related tests and discuss the main findings from the statistical analysis.

Step 4 (Session 3b)

Relate the statistical findings to the theoretical framework, discuss the results with students in groups and write the main conclusions derived from the case study.

Case

Step 1: Theoretical framework

Complaints in the tourism industry have always existed, but these were always off-line, with data being costly to gather and process. The emergence of *online complaints* has paved the way for receiving instantaneous feedback from customers and, as this is often anonymous, it appears to be more reliable and useful. The opinions of clients after their visits usually include key information on business weaknesses and allow hotel managers to improve loyalty relationships with their clients. Different techniques based on text analysis of online reviews have been used in the hotel industry, including data mining, opinion mining, and semantic and sentiment analysis. Part of the methodologies employed consisted of identifying words or expressions in reviews on different aspects of the service or facilities. The most repeated words were then identified as those corresponding to customers' major concerns. Obtaining massive volumes of information immediately and for free has been of great advantage to researchers and managers since the emergence of online tourism platforms.

There are about one thousand amusement and theme parks in the USA and Europe. The various Disney Parks around the world were visited by approximately 156 million people in 2019, while the theme park industry in Europe received 72 million visitors, showing the importance of theme parks as tourist attractions. Although online reviews in the tourism industry started by focusing on the performance of hotels, this marketing tool has gained importance for theme parks in recent years. Managerial staff focus on meeting customer expectations, while also raising their level of satisfaction, by monitoring and adopting a continuous customer experience management policy (Banerjee & Chua, 2016; Berezina et al., 2016). The conditions of the park experience, including crowds, queues and the quality of the facilities, rank first in these reviews, followed by pricing issues. TripAdvisor is a very important platform in this area, appearing as the main source of information for this type of research. More than 700 million reviews and comments are posted annually on this tourism and travel platform, which also allows people to post reviews on theme parks.

The content generated by internet users can be considered as a mix of facts, opinions, feelings and emotions. These feelings and emotions affect consumer experience and satisfaction levels. In particular, there are several factors that influence the *perceived experiential value* of visitors to theme parks. The four primary dimensions of experiential quality value are interaction, physical environment, general outcome and access conditions (Wu, Li & Li, 2018). Results suggest that the physical environment appears to be a key component of customer experience. However, access conditions also play a key role in influencing reviews. For example, two main issues emerge in this area: (1) paid parking (with no free parking option available); and (2) rules against entering the parks with food and drink. These regulations entail extra expenses for visitors to the theme parks, and often result in customers complaining in their online reviews.

Step 2: Downloading data from the online platform and preparing the dataset

Theme parks have been described as forms of entertainment for many children and adults. However, as in the case of any entrepreneurial activity, complaints arise regarding different aspects that create dissatisfaction for visitors. Throughout this case study, we will focus on two specific complaints that often appear in the reviews of theme parks in Spain:

1. *Outside Food Policy (OFP)*: Visitors are not allowed to enter most theme parks with their own food and drink, and often end up spending additional money in the parks' restaurants, which tend to be more expensive than restaurants outside of the parks. Additionally, almost all theme park restaurants only serve fast food and, as a result, the cost of the visit increases but the quality of the meals decreases.
2. *Paid Parking (PP)*: In most theme parks, the car parks are not free, adding an extra cost to the visit. Visitors may decide to park outside the theme park area, but this means they must often walk a considerable distance to reach the park entrance.

The tourism platform TripAdvisor.com plays an important role by offering real-time information on customers' opinions in online reviews of the industry. In this study, we focus on reviews of the two most popular theme parks in Spain, located close to the two biggest cities in the country. PortAventura in Salou, a coastal destination close to Barcelona, was the sixth most visited theme park in Europe in 2019, with 3.7 million visitors. Parque Warner Madrid occupied the twelfth position in the European ranking, with 2.2 million visitors. Both parks follow similar rules in terms of the PP and OFP issues analysed here. Parking at PortAventura normally costs €12 per day but is free for visitors staying at one of the park's hotels, while at Parque Warner Madrid, the price of open and covered parking facilities is €10 and €12, respectively. This information can be found on the parks' websites, but it does not appear on their general price lists.

In practice, this means that all visitors arriving in their own vehicles must pay an additional cost. It should also be noted that Parque Warner Madrid is very far from the nearest urban centre, San Martín de la Vega, so it is not possible to park outside the premises and then walk to the entrance. However, PortAventura is adjacent to the outer part of the urban centre of Salou, which allows visitors to park for free in the streets. This possibility is mentioned in the comments section of TripAdvisor, but it also states that this involves a 10–15-minute walk to the park entrance. Moreover, the number of available parking spaces in nearby public roads is very limited in the low season, and almost non-existent in the high season.

In this section, after gathering data from the TripAdvisor portal, we check whether customer complaints follow a similar distribution of differences in the case of the two selected parks. On the 30th June 2018, the day we gathered the data,

TripAdvisor recorded a total of 4,780 reviews for Parque Warner Madrid (87% in Spanish), and 19,850 for PortAventura (54% in Spanish). From this raw data stock downloaded directly from the website, we selected those reviews that, from the 1st January 2017 on, made any reference to the two main issues mentioned. In total, we collected 1,019 opinions for Parque Warner Madrid and 1,377 for PortAventura on the Outdoor Food Policy (OFP) and Paid Parking (PP) issues.

To do so, we employed the keyword search engine available on the TripAdvisor portal, both in Spanish and in English. After this filtering procedure, it was necessary to carry out a manual review to clean up the downloaded records. For example, when identifying the reviews and opinions about the Paid Parking (PP) issue, we employed three main keywords: parking/aparcamiento/aparcar (in Spanish), which meant we had to perform three different searches, one for each word. We also checked the opinions which directly referred to the rule against bringing food and beverages (OFP) inside the park. We then performed a final check of the dataset to make sure of its complete validity. This was important as there were reviews which contained the keywords, but which were not directly related to the focus of the current analysis. For example, some online reviews mentioned the length of the queue to enter the car park, or the long distance from the car park to the park entrance.

We also added around 30% of additional reviews which did not contain complaints about these two issues specifically, in order to avoid selection bias in the statistical test to be employed in this section. After the data cleansing process, we finally selected 586 negative reviews directly referring to the PP and OFP issues, 333 for Parque Warner Madrid and 253 for PortAventura.

Steps 3 and 4: Running the statistical test on the data and reaching the main conclusions from the analysis

Once we had collected our dataset, we loaded it into a statistical software program for the subsequent analysis. There were several programs that could have helped us in the statistical procedure, including a simple Excel spreadsheet, or more complex software such as the SPSS (IBM), EViews (QMS) and STATA (StataCorp) packages. In our case, we used the STATA software, and started performing a descriptive analysis of the data.

Our dataset included three types of variables for the reviews downloaded from the TripAdvisor website:

1. INFO: The review contains no express complaints about either the PP or OFP issues, but one or both are mentioned in the text. We labelled these as 'informational references'.
2. COMP: The review makes an express complaint about the PP or OFP issues. We labelled these as 'express complaints'.
3. REVIEW: The review includes either informational references (INFO) and/or express complaints (COMP) about the PP and OFP issues.

The structure of the dataset showed that, in descriptive terms, the informational references (INFO) accounted for 22% of the total reviews in the sample data for the case of PortAventura (8% for PP and 14% for OFP), and 29% in the case of Parque Warner Madrid (13% for PP and 16% for OFP). In the case of express complaints (COMP), PortAventura represented 22% of the total reviews in the sample concerning the two issues under study (11% for PP and OFP, respectively), while Parque Warner Madrid made up 29% (16% for PP and 13% for OFP). Finally, regarding the REVIEW variable, total reviews (INFO + COMPL) mentioning both PP and OFP in the data sample accounted for 44% for PortAventura (19% for PP and 25% for OFP), and 58% for Parque Warner Madrid (29% for both PP and OFP).

As shown, descriptive measures seemed to point to existing differences between both parks in terms of the relative weight of the three types of variables under analysis. Parque Warner Madrid seemed to receive around 30% more informational references (INFO) and express complaints (COMP) of the total reviews on the PP and OFP questions than PortAventura. Moreover, the informational references (INFO) seemed to show a higher number of mentions of OFP for both parks, while for express complaints (COMP) they appeared to be more important in relative terms for the PP issue. At this point, and to get a deeper understanding of the potential differences arising between online customer behaviour for the two parks, we ran an Analysis of Variance (ANOVA) statistical test. This allowed us to compare the whole distributions of the three variables for the two parks, in search of what we call a 'park effect'.

ANOVA analysis: In search of a 'park effect' in the INFO-COMP-REVIEW variables

The Analysis of Variance (ANOVA) procedure allowed us to identify *from a statistical point of view* whether two independent distributions behaved similarly or not.[1] Furthermore, the one-way ANOVA procedure allowed us to test if any single explanatory variable (predictor variable) could have influenced the observed distributions under study (outcome variables). In this case, the analysis focused on identifying potential differences arising between both theme parks regarding the post-visit opinions of customers, adding new information to that of the purely descriptive approach. In general, the ANOVA test compared two distributional values (online reviews) taken from two independent groups (parks). The null hypothesis (H0) to be tested was that the distributions of the reviews would not differ statistically between the two parks. H0 would therefore be rejected if differences were shown to exist between the two parks. The one-way ANOVA procedure allowed us to then test if the distribution of reviews statistically differed between the two parks.

Table 5.1 summarises the output of the ANOVA procedure applied to the visitors' reviews, while defining a (dummy) park ID variable (PortAventura=0, Parque Warner Madrid=1) as the control or predictor variable. In this regard, the results are explained for the three variables defined in the analysis, namely the INFO, COMP and REVIEW variables. Table 5.1 contains the ANOVA analysis for the PP issue in the first place, followed by the OFP issue, and then PP and

TABLE 5.1 ANOVA procedure for the three defined sets of variables

ANOVA INFO_PP ParkID
Number of obs = 586 R-squared = 0.0068
Root MSE = 0.3048 Adj R-squared = 0.0051

Source	*Partial SS*	*df*	*MS*	*F*	*Prob > F*
Model	0.3743	1	0.3743	4.03	**0.0452**
Park ID	0.3743	1	0.3743	4.03	**0.0452**
Residual	54.2758	584	0.0929		
Total	54.6501	585	0 .0934		

ANOVA COMP_PP ParkID
Number of obs = 586 R-squared = 0.0040
Number of obs = 586 R-squared = 0.0040

Source	*Partial SS*	*df*	*MS*	*F*	*Prob > F*
Model	0.2851	1	0.2851	2.37	**0.1242**
Park ID	0.2851	1	0.2851	2.37	**0.1242**
Residual	70–24	584	0.1202		
Total	70–52	585	0.1205		

ANOVA REVIEW_PP ParkID
Number of obs = 586 R-squared = 0.0121
Root MSE= 0.4276 Adj R-squared = 0.0105

Source	*Partial SS*	*df*	*MS*	*F*	*Prob > F*
Model	1.3129	1	1.3129	7.18	**0.0076**
Park ID	1.3129	1	1.3129	7.18	**0.0076**
Residual	106.79	584	0.1828		
Total	108.10	585	0.1847		

ANOVA INFO_FF ParkID
Number of obs = 586 R-squared = 0.0008
Root MSE= 0.3577 Adj R-squared = 0.00065

Source	*Partial SS*	*df*	*MS*	*F*	*Prob > F*
Model	0.0623	1	0.0623	0.49	**0.4855**
Park ID	0.0623	1	0.0623	0.49	**0.4855**
Residual	74.7226	584	0.1279		
Total	74.7849	585	0.1278		

ANOVA COMP_FF ParkID
Number of obs = 586 R-squared = 0.0020
Root MSE= 0.3245 Adj R-squared = 0.0003

Source	*Partial SS*	*df*	*MS*	*F*	*Prob > F*
Model	0.1239	1	0.1239	1.18	**0.2784**
Park ID	0.1239	1	0.1239	1.18	**0.2784**
Residual	61.51	584	0.1053		
Total	61.63	585	0.1053		

(continued)

TABLE 5.1 Continued

ANOVA REVIEW_FF ParkID
Number of obs = 586 R-squared = 0.0031
Root MSE= 0.4438 Adj R-squared = 0.0014

Source	*Partial SS*	*df*	*MS*	*F*	*Prob > F*
Model	0.3620	1	0.3620	1.84	**0.1757**
Park ID	0.3620	1	0.3620	1.84	**0.1757**
Residual	115.03	584	0.1969		
Total	115.39	585	0.1972		

ANOVA INFO_PP_FF ParkID
Number of obs = 586 R-squared = 0.0040
Root MSE= 0.3468 Adj R-squared = 0.0023

Source	*Partial SS*	*df*	*MS*	*F*	*Prob > F*
Model	0.7421	1	0.7421	3.93	**0.0480**
Park ID	0.7421	1	0.7421	3.93	**0.0480**
Residual	110.37	584	0.1889		
Total	111.11	585	0.1899		

ANOVA COMP_PP_FF ParkID
Number of obs = 586 R-squared = 0.0070
Root MSE= 0.4375 Adj R-squared = 0.0053

Source	*Partial SS*	*df*	*MS*	*F*	*Prob > F*
Model	0.7851	1	0.7851	4.10	**0.0433**
Park ID	0.7851	1	0.7851	4.10	**0.0433**
Residual	111.78	584	0.1914		
Total	112.57	585	0.1924		

ANOVA REVIEW_PP_FF ParkID
Number of obs = 586 R-squared = 0.0209
Root MSE= 0.4954 Adj R-squared = 0.0192

Source	*Partial SS*	*df*	*MS*	*F*	*Prob > F*
Model	3.0539	1	3.0539	12.44	**0.0005**
Park ID	3.0539	1	3.0539	12.44	**0.0005**
Residual	143.33	584	0.2454		
Total	146.39	585	0.2502		

OFP altogether. The first table of each block of three tables contains the results of the ANOVA test for the INFO variable, the second table of each block contains the results for the COMP variable, and the third table contains the results for the REVIEW (INFO + COMP) variable.

The first block in Table 5.1 contains three tables. The output of the ANOVA procedure includes the result of the test for the (full) model that matched the

ParkID model, as we only had one control variable in this case (ParkID). For this reason, the Partial SS results for the model and ParkID become the same. In this way, each statistical output shows data for the Sum of Squares (SS) results, degrees of freedom for the test (df), the corresponding F-test value, and the probability of rejection of the H0 hypothesis of equality of distributions for the two parks, PortAventura and Parque Warner Madrid. The three tables analyse the INFO variable (PP, OFP, and PP+OFP). We can see that the distributions differ statistically in the PP and PP+OFP reviews, with a p-value below the 5% threshold, i.e., we can reject the H0 of equality with a 95% significance. However, we cannot reject the equality hypothesis for the OFP issue in the second row of the table, with a p-value of 48%. In this case, we can state that, statistically, the distribution of informational references (INFO) of online customer reviews differed for paid parking (PP), but not for the outside food policy (OFP). In the third row, including the joint analysis of PP+OFP, we also see the presence of a 'park effect'.

In the case of the 'express complaints' (COMP) variable, the three tables in Table 5.1 show that we can identify a park effect in the case of PP, at 88% significance, and for the joint PP+OFP issues, but not for the single OFP issue in the second table of the panel (prob=27%). Once more, the online reviewing behaviour of visitors to these two theme parks differed clearly when writing express complaints about the Paid Parking (PP) issue, but not for the Outside Food Policy (OFP).

Finally, in the case of informational plus complaint references of visitors through the TripAdvisor website (the REVIEW variable), the statistical analysis of the remaining block in Table 5.1 shows the existence of differences between the parks, which we call a 'park effect', in the case of the PP and PP+OFP issues, but not for OFP (prob=17%).

Conclusion

Summarising the results of this case study, we compared the two most popular theme parks in Spain, PortAventura and Parque Warner Madrid, seeking a better understanding of the factors which influence visitor satisfaction with the whole experience, as expressed in their online reviews at one of the top travel and tourism digital platforms in the world, TripAdvisor.com. In order to do so, the exercise began by introducing two new issues related to access conditions as defined by the tourism marketing literature, and which had not been previously researched. As reflected in the gathered online reviews, two issues were of significant importance for customers regarding their visits to the parks; namely, having to pay to use the car park, and rules against bringing food and beverages inside the parks.

In a first descriptive approach, the data showed some traces of potential existing relative differences in the reviews made by visitors concerning these two main issues. We then applied a deeper analysis employing statistical methods to better ascertain such issues. The ANOVA test showed the existence of what we have called a 'park effect', or statistical distributional differences between the reviews of customers visiting both parks, clearly arising in the case of Paid Parking (PP), but not so for the Outside Food Policy (OFP), despite both PP and OFP issues being mentioned by visitors in their reviews, either informationally or in express complaints.

Managerial implications emerging from the case study include the fact that the paid parking issue seems to be of higher concern mainly for visitors to Parque Warner Madrid, so managers should look into this issue as it affects accessibility to the park. However, outdoor food policies seem to be having a similar effect on both PortAventura and Parque Warner Madrid, given that visitors to the two parks include this issue in their online reviews with similar frequency, as shown by the ANOVA statistical analysis, either informationally or through express complaints.

Note

1 See, for example, https://statistics.laerd.com/stata-tutorials/two-way-anova-using-stata.php for a detailed explanation of the ANOVA test and STATA related procedure.

References

Banerjee, S., & Chua, A.Y. K. (2016). In search of patterns among travellers' hotel ratings in TripAdvisor. *Tourism Management*, 53, 125–131.

Berezina, K., Bilgihan, A., Cobanoglu, C., & Okumus, F. (2016). Understanding satisfied and dissatisfied hotel customers: Text mining of online hotel reviews. *Journal of Hospitality Marketing & Management*, 25(1), 1–24.

Wu, H.-C., Li, M.-Y., & Li, T. (2018). A study of experiential quality, experiential value, experiential satisfaction, theme park image, and revisit intention. *Journal of Hospitality & Tourism Research*, 42(1), 26–73.

Further reading

Kotler, P., Bowen, J. T., Makens, J., & Baloglu, S. (2017). *Marketing for Hospitality and Tourism*. India: Pearson.

Mellinas, J. P., & Reino, S. (2019). eWOM: the importance of reviews and ratings in tourism marketing. In M.A. Camilleri (Ed.), *Strategic Perspectives in Destination Marketing* (pp. 143–173). Hersey, PA: IGI Global.

Case 6

ANALYSING INFORMATION ABOUT HOTELS ON ONLINE REVIEW SITES UNDER SPECIAL CONDITIONS

Juan Pedro Mellinas, Andrés Artal-Tur and Noelia Sánchez-Casado

Duration

2.5 hours.

Learning objectives

Upon completing the case, participants will be able to:

- acquire a basic knowledge about the use of online reviews in research
- download and describe a dataset retrieved from an online review site
- learn about the implied statistical effects of a sample selection bias.

Target audience

Undergraduate and postgraduate students, who are required to possess a basic knowledge of descriptive statistical analysis and its interpretation (e.g., rate, average, median, mode, minimum, maximum, frequency distribution scale), together with the concept of 'bias'. The case study is suitable for individual or group work.

Teaching methods and equipment

The case study uses hands-on practice and brainstorming as its teaching methods. Students are required to be able to retrieve data from an online review site and use a Microsoft Excel spreadsheet to analyse it. Then, if the classroom is equipped with computers and an internet connection, the students could start working on the case study during the lesson, although no specific equipment is actually required.

DOI: 10.4324/9781003182856-6

Teaching instructions

The students should read the case study text and then elaborate reasoned answers to the questions included at the end. It would also be interesting to encourage discussion during the class, and for students to hold a brainstorming session.

At the end of the exercise, instructors can help to summarise and form a final block of learning results and conclusions emerging from the case study.

Instructors should follow the below steps during the development of the case study.

Step 1

Help students identify the clear bias that can occur when using two samples that do not include the same type of hotels and which have been compared as if they had similar characteristics.

Step 2

Once the bias problem has been identified, help the students to find the most optimal solution to avoid bias. In this case, it is advisable to use a sample including hotels that remained open during the same two-year period.

Step 3

Show the students the analysis of what happened to luxury hotels in a large European capital during the pandemic period (after March 2020) using reviews on TripAdvisor, Expedia and Booking.com, identifying:

- the number of reviews and the average scores before and after March 2020
- the number of weeks without any reviews (forced closure period) from March 2020 to March 2021.

The number of reviews (85,000) during the first analysed period, prior to the pandemic, is for the total number of hotels in the urban destination, so the first score is obtained from the real statistical hotel population. However, the number of reviews (15,000) gathered during the second analysed period, during the pandemic, only represents the number of hotels that stayed open.

If hotels had been closed randomly during lockdown, rather than being forced to do so by their own particular financial conditions, the differences in the average scores for these two periods would have reflected only the changes which occurred in levels of customer satisfaction. However, as observed during lockdown, luxury hotels were forced to close much sooner compared to mid-range and basic hotels (Lucente Sterling, 2021). In particular, given the existence of such sample selection bias, the ratings and scores in the gathered data sample for hotels changed accordingly; this being the idea we will develop throughout this case study exercise. As a result, we will highlight how the lower ratings and scores presented throughout the confinement period for hotels were affected by a sample selection bias effect, which in fact can help to shed more light on the issue for any interested students.

Furthermore, not only the hotels but also segments of customers staying at the remaining hotels may also have changed during these two periods. In other words, the purpose of the trip may have changed during the pandemic (i.e., business and health vs holiday trips), together with the products and services available (i.e., breakfast, spa, gym and swimming pool services), or the size of the party (i.e., travelling alone vs doing so with a partner, friends, or family). All these issues would also affect the scores posted after the visit.

In this context, and to avoid any sample selection bias resulting from the pandemic situation, the average score must be recalculated, considering only those hotels which remained open during both periods, hence deleting from the sample those hotels which were closed during the pandemic. By applying this proposed correction to the data, the remaining changes observed in the hotel scores would capture the existing changes in customer perception, without any bias induced by any kind of statistical artifact.

Each student will be provided with an Excel template, which includes the names of the variables to be gathered, and will then be assigned two to three hotels from a specific destination, so the teacher must previously define a sample and provide the names of some of these specific hotels to the students.

The students must then access an online review site, search for their assigned hotels, sort the reviews by date and perform a manual data collection for the variables included in the Excel template. At the end of the process, the data from all the students is merged into a single shared database in order to carry out the data analysis.

Step 4

Students will be able to determine whether the two following hypotheses are true or not:

1. A significant number of luxury hotels were closed during the toughest year of the pandemic.
2. Customer satisfaction dropped significantly during the pandemic.

A similar exercise can be carried out for destinations where we know there is a strong seasonality, with many hotels closing during the low season.

Case

Theoretical framework

Nowadays, most travellers around the world consult online review sites about destinations, hotels and attractions in order to plan their holiday trips. Online hotel reviews have become a fundamental source of information not only for potential guests, but also for hotel managers interested in opinions about their services and the quality of their personnel, as well as for academics carrying out new research into the tourism marketing area.

The existence of millions of hotel reviews provides valuable information for learning about customer behaviour, tastes, complaints and new consumer trends. Demand segmentation analysis allows firms to identify the preferences and needs of customers, while designing and implementing new strategies for targeted segments, which in turn results in an increase in customer satisfaction levels and related revenues.

In this context, segmentation analysis becomes an effective marketing tool for planning a strategical approach, helping to classify customers according to their particular requests, characteristics, preferences and behaviours. Customers may come with unique service requests that are often difficult to provide, a situation in which segmentation analysis has proven to be helpful for business management. It helps to group similar customers, identify preferred target markets and provide information about marketing mix elements to ensure effective marketing policies. Moreover, segmentation approaches in tourism contexts emphasise the determination of requests and profits related to specific product groups, purchasing/usage behaviours, demographics and nationality. However, the choice of segmentation approach has become more complex in times of rapid technological development and big data opportunities, allowing for data mining and machine learning methodologies.

Whenever applicable, researchers are replacing information traditionally collected through paper questionnaires and interviews with that gathered from online review sites, such as Booking.com, with 205 million hotel reviews in 2019, and TripAdvisor, with 900 million reviews on tourism businesses and destinations in the same year. Online datasets have proven to be quite helpful for extracting information on hundreds of hotels from thousands of reviews. By using any web data extraction software, the process can be automated, allowing for new sets of customer information on thousands of hotels to be taken from millions of reviews on a monthly, or even weekly, basis.

A medium size hotel can easily receive more than one thousand reviews from different websites, and new reviews are uploaded every month. In the case of large hotels, the number of reviews becomes huge, and even difficult to manage. As an example, the Circus Casino Hotel in Las Vegas (3,774 rooms) received around 27,000 reviews on Expedia in 2019, 14,000 on TripAdvisor, 24,000 on Booking.com and 64,000 on Google platforms.

Hotel reviews allow us to produce a global score of customer perception of the hotel services received, together with single scores and written opinions of customers regarding their stay. Figure 6.1 shows global information on the accommodation services and customer reviews appearing on the TripAdvisor website.
It shows certain characteristics of the visitors and their reviews, such as helpfulness votes, geographical origin or link to visitor profile, together with the review of the stay, including aspects such as the visiting dates or trip type.

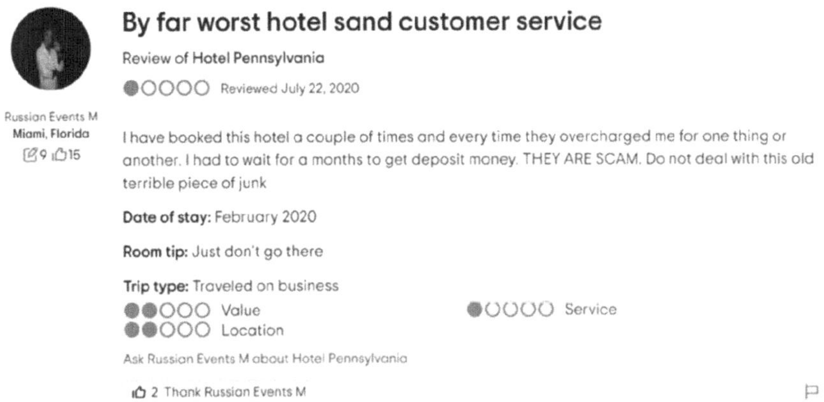

FIGURE 6.1 Review on TripAdvisor

Source: TripAdvisor (https://www.tripadvisor.com/ShowUserReviews-g60763-d214197-r761109129-Hotel_Pennsylvania-New_York_City_New_York.html)

TripAdvisor allows people to customise searches by calendar quarters. In addition, each review includes the date when it was written, and the date of the guest's visit to the hotel. This allows, when downloading reviews and creating a dataset, for opinions to be sorted or filtered by date. It then helps to examine whether the level of customer satisfaction is influenced by the season of the visit, and to check if any changes in the hotel context for a specific date could affect the level of declared customer satisfaction.

The Covid-19 pandemic hit the tourism and hospitality industry hard, dramatically reducing revenues and the number of visitors at most of the world's destinations. The bulk of hotels found themselves in a difficult situation from an entrepreneurial perspective, given the sharp decrease in the number of visitors and related revenues which occurred, and the temporary or permanent firing of employees. Occupancy rates fell by between 30% and 50%, while international flights were cancelled and only limited domestic trips were allowed during the summer. This situation is somewhat similar to that occurring at highly seasonal destinations, but in this case such a sharp drop in demand and accessibility conditions could not have been anticipated by hotel and destination managers.

Moreover, many of the hotels that remained open had to lean on the financial support offered by regional and national governments. For several months, most accommodated guests were either health care workers or homeless people suffering the economic consequences of the pandemic. Thanks to these government actions, a significant number of hotels managed to survive and keep their establishments open, despite the significantly reduced number of conventional guests arriving. During this difficult situation, the main segment of hotels receiving public aid was that of low to mid-range hotels with adjusted room fees. This was therefore the segment with the greatest level of access to public contracts with the health services, social assistance, the army and local governments.

On the other hand, the luxury hotel segment did not benefit from this type of public aid. Moreover, people who travelled during this period (from March 2020 to February 2021) did so mainly for business and not for leisure. This brought about a significant reduction in tourism revenues at major destinations, with hardly any guests staying at luxury hotels during the lockdown period.

Data analysis

To better understand whether the pandemic situation may have influenced the related level of visitor satisfaction either positively or negatively, all the types of accommodation included in the 'hotel' category in a large urban destination were considered as our initial data sample. The results of the search included a significant number of 6- and 5-star hotels, plus a larger number of luxury 4-star hotels, with the average hotel having more than 200 rooms and being well situated in the downtown historical centres of the urban destinations. As mentioned previously, most of them were closed during the March–April 2020 confinement period, and by March 2021 many still remained closed.

To measure the level of customer satisfaction with their stay, we employed the hotel score rating value on Booking.com. It is interesting to note that more than 50% of rooms are booked through an online platform, with some 40% of clients doing so through Booking.com. These clients may later respond to a survey on the same website aimed at rating their experience of the hotel stay. This ensures that the data obtained from Booking.com is reliable for measuring the satisfaction of hotel guests.

Using specialised data software, we downloaded all the reviews posted on Booking.com for hotels in the urban destination during March 2021. We know that Booking.com only keeps the reviews posted over the last two years in order to ensure up-to-date information on the hotels included in their database. Therefore, the reviews obtained were from March 2019 to March 2021. It is worth noting that a number of these reviews, those made from March 2019 to February 2020, included those stays not hit by the pandemic, while the rest, those made from March 2020 to March 2021, showed the effect of the pandemic.

A total of 100,000 reviews were obtained for the urban destination for this two-year period. The data for each of the downloaded reviews included the name of the hotel, the date when the review was made, and the average score given by the visitor. Once all the data had been gathered from the website, we broke the initial sample down into two groups: (1) reviews posted prior to the pandemic and (2) reviews posted after the outbreak of the pandemic, starting in March 2020. For both groups, we computed the average Booking.com score, as shown in Table 6.1.

As shown in Table 6.1, there exist important differences between the two periods under analysis, with scores declining from the initial 8.1 points to 7.5 for the second

TABLE 6.1 Data from Hotels at Urban Destination 'X' from www.booking.com

Period of Analysis	*Number of Reviews*	*Average Booking.com Score*
March 2019 – February 2020	85,000	8.1
March 2020 – February 2021	15,000	7.5

Source: Own elaboration from retrieved dataset www.booking.com

period. The first noteworthy point is that, although both periods are from one year, there were far fewer reviews during the pandemic period, which is totally logical and to be expected due to hotel closures and drops in occupancy rates. Moreover, as shown in Table 6.1, the average score of the hotels dropped by 0.6 points from one period to another, which seems to indicate a significant drop in the satisfaction of hotel customers due to the new situation brought about by the COVID pandemic.

Open questions for students at the end of the session

1. Do you agree with the last statement?
2. Are there any other conditions that could help to explain the lower level of satisfaction observed during the pandemic period?
3. What information would you need to check for any bias in the data analysis?
4. If there were any problems in the dataset, what tools and information would you need to fix them?

References

Lucente Sterling, A. (2021). What Happens to All of NYC's Empty Hotels? *www.ny1.com/nyc/all-boroughs/news/2021/01/15/what-happens-to-all-of-nyc-s-empty-hotels*

Further reading

Gursoy, D. (Ed.). (2018). *The Routledge Handbook of Hospitality Marketing*. Abingdon: Routledge.

Case 7

EFFECTIVE AIRLINE MARKET SEGMENTATION

The case of Singapore Airlines Group

Evi Chatzopoulou, Ioulia Poulaki and Andreas Papatheodorou

Duration

Regardless of the teaching method used, the instructor should encourage students to prepare a scenario for resolving the case study issues. The students study the text, analyse it and identify the key issues. It is recommended that students should be given 25–30 minutes to read the case study, and 45 minutes to prepare the presentation about the outcomes by using the hands-on discussion method. In the case of a homework assignment, students should be given 20–25 minutes to read the case study and up to 15 minutes to pose questions to the lecturer. Then students may be assigned a short description of the solution as homework and devote 30 minutes for presentation and discussion in the following lecture session. Finally, 10 minutes may be devoted for summarising the outcomes based on marketing theory.

Learning objectives

Upon completing the case, participants will be able to:

- acquire a critical and reflective understanding of STP (Segmentation, Targeting, Positioning)
- develop a more critical appreciation of segmentation within the marketing process
- critically assess the effectiveness of different segmentation approaches in relation to various airlines from a strategic perspective
- understand the particularities of the airline industry in terms of customer service planning according to their target group
- structure customer persona profiles for airlines and other tourism businesses.

DOI: 10.4324/9781003182856-7

Target audience

This case study may be of added value to both undergraduate and postgraduate programmes. In particular, the case study is predominantly suitable for modules in tourism marketing, marketing fundamentals and strategic marketing. This case study can be discussed in small student groups.

Teaching methods and equipment

This case study contributes into the development of critical thinking of students regarding alternative ways to approach marketing issues faced by the Singapore Airlines (SIA) Group. This process facilitates the understanding of SIA Group's complexities and problems by students who can provide different yet acceptable approaches. Furthermore, this case study allows the instructor to focus on the students' level of understanding and form the proper working groups to achieve the optimal quality of deliverables. The working pace of an individual is as important as the level of their understanding and, thus, the instructor should achieve a group mix that supports all students in a way that the production of ideas and proposed solutions are generated based on collective effort. This way the instructor can help students expand their knowledge considering the personal skills, abilities and qualities of everyone within the group.

Although customer segmentation based on services is an important part of this case study, the potential to apply targeting and positioning strategy is also significant. It is suggested that the instructor should start by analysing segmentation issues and later emphasise customer personas, targeting and positioning. Prior to the case study project discussion, the introduction of the case should take place in the classroom where a hands-on method should be applied by the instructor to answer questions and/or to build-up the introduction process. The case study can be discussed in small student groups, who are also encouraged to build customer personas. It is strongly recommended that the instructor should ensure that the discussion at the end of the case study becomes a mandatory as part of the learning process, with emphasis on generalisation and wider applicability.

Teaching instructions

Students need to familiarise themselves with the case before discussing it. First, the case study's goals should be briefly introduced by the instructor. The discussion may start with the evaluation of the air transport market, the penetration of Low-Cost Carriers (LCCs), and the challenges faced by air transport in the post Covid-19 world and how SIA Group may deal with its competitors and customers. Following this, customer persona profiles can be presented in PowerPoint slides in such a way that the features of each profile can be gradually identified by the students. Moreover, at the end of the present case study, there are seven questions for the students with indicative answers as follows.

The first question refers to segmentation criteria. SIA Group identifies several ways to segment the market and develops profiles of the resulting market segments. It uses multiple segmentation bases to identify smaller, better-defined target groups of consumers. For leisure consumers, the major segmentation variables are geographic, demographic, psychographic and behavioural. Business buyers can be segmented by using geographic, demographic, benefits sought, user status, usage rate and loyalty status segmentation. Business buyers are also segmented based on their purchasing approach, situational factors and personal characteristics. Scoot, SIA Group's low cost subsidiary, uses mainly demographic and psychographic segmentation criteria based on its competitive advantage due to its unique specialisation in low-cost, long-haul services.

For the second question, students should focus on special services and passenger treatment during the COVID-19 pandemic era. A customer persona should include a customer's bio, pain points, profile info, communication methods, photo and a random name of the target passenger.

The third question refers to major strategies that are generally used to target market segments. SIA has adopted differentiated marketing, based on different market offers for the various segments, whereas Scoot has adopted concentrated marketing, focusing on only one, or a few, market segments. The targeting strategy depends on company resources, product variability, product life-cycle stage, market variability and competitive marketing strategies.

Regarding the fourth question, students should analyse the effectiveness of market segmentation based on theory. To target the best market segments, a company should first evaluate each segment's size and growth characteristics, structural attractiveness and compatibility with company objectives and resources.

The fifth question is about the different types of value propositions around which a company can position its products. Students should recognise the unique characteristics that give SIA Group airlines their competitive advantage. SIA focuses on services whereas Scoot focuses on price differentiation.

For the sixth question, students should explain the importance of positioning for competitive advantage. In planning their differentiation and positioning strategies, students should prepare perceptual positioning maps that show consumer perceptions of SIA Group brands vis-à-vis competing products on important buying dimensions. Students can use key characteristics such as services–price, regions–services, regions–price.

The seventh question is based on the previous one. Students should explain the SIA Group airlines' position strategies according to the perceptual position maps. The number and strength of competitors will shape potential opportunities.

The students may split into groups of two to four individuals. It is important that all students get a chance to share and challenge their opinions. Each group may be asked to discuss key issues, consider and justify potential solutions, and prepare a short presentation of the outcomes. Students may write a short description of their outcomes and upload it on the e-learning platform of their institution and/or discuss it in class. Further to the above, through this airline case study, the students are requested to use material developed by SIA and Scoot to build new customer persona profiles based upon the changes that emerged after the Covid-19 pandemic outbreak. All the students' submissions should be discussed by the instructor questing for an appropriate (optimal) solution to the SIA problems. Each student group leader may be asked to present the team's output in front of the whole class, suggesting an appropriate way forward. Questions may be posed from the instructor or other students. At the end of each session, the instructor may summarise the case and the outcomes of the students' groupwork.

Case

Introduction

Airline-within-Airline strategy has been developed at an important level in the airline industry, aiming to address distinct target groups. Multi-brand strategies have emerged because of intense competition and market saturation. Singapore Airlines (SIA) Group implements these strategies to appeal to new customer segments, thus remaining competitive in the aviation industry. Market segmentation refers to the customer profile-building process into segments where the included profiles in each of the segments present similar characteristics and interests. The SIA Group invests in both full-service and low-cost airlines, targeting passengers who travel for business or pleasure, thus meeting the demand of most market segments. The SIA Group's portfolio comprises Singapore Airlines, Scoot, SilkAir (now absorbed by Singapore Airlines) and Singapore Airlines Cargo which operates a total fleet of 196 (122, 49, 25 and 7 respectively) aircraft. Prior to the Covid-19 pandemic, the SIA Group used to serve 136 destinations, making it one of the diverse organisations in the Asia Pacific. Due to the Covid-19 pandemic travel restrictions, it had been operating a reduced network in 2020 and 2021 (Scoot Airlines website, 2021; Singapore Airlines website, 2021a, 2021b).

Singapore Airlines, operating as a full-service network carrier (FSNC), focuses on the full-service passenger segment serving both short- and long-haul destinations. SilkAir provided passenger air transportation with a focus on the full-service passenger segment serving regional markets. However, SilkAir was folded into SIA as a wholly owned subsidiary of Singapore Airlines at the end of January 2021. Thus, SIA has become responsible for SilkAir managerial operations and online presence. SilkAir has upgraded its services, adopting SIA's practices and features, and offering the same in-flight and ground services. This strategic decision is aimed at enhancing the SIA Group's short and medium-haul destinations with connections to more exotic destinations. Scoot provides passenger air transportation, with a focus on the low-cost passenger segment. Heavy competition and the evolution of Low-Cost Carriers (LCCs) over time, led the SIA Group management to decide to launch a low-cost product into the long-haul markets. In the following, the customer service of Singapore Airlines and Scoot will be discussed.

Singapore Airlines

Singapore Airlines (SIA) is the flagship carrier of Singapore, based at Singapore Changi Airport. Over sixty routes were operated by Singapore Airlines in early 2021, a reduction of approximately 50% compared to the pre-Covid-19 period. Being a premium brand, SIA targets wealthier customers who value comfort and service excellence, and is attractive to business travellers. Passengers can choose among three types of fare, *lite*, *standard* and *flex*, characterised by different allowances and restrictions. As a result, SIA actively seeks effective differentiation (Singapore Airlines website, 2021).

SIA is regarded as a pioneer in emphasising customer service in the aviation industry. SIA is one of the most awarded airlines worldwide for its continuous efforts to maintain service standards both in-flight and on the ground. It was awarded the 'Best International Airline' by Travel & Leisure in 2020, and the 'Best Airline for Business Travel', 'Best First Class', and 'Best Economy Class' by Business Traveller in the same year. In the preceding years it won awards in several categories such as 'Best Long-haul Airline' (6 times), 'Best Airline in the World' (31 out of 33 years), 'Best Seat Comfort and Best In-Flight Entertainment for the Eastern Asia Region', and 'Best Airline for Business Travellers Worldwide' (19 times). SIA topped the list of international airlines in the categories of safety, trust, customer service and food.

Being a city-state, Singapore lacks a domestic aviation market. Thus, SIA had to seek an effective differentiation strategy at an international level since its inception. It created a marketing icon – the Singapore Girl in 1970s. The image of the 35-year-old (timeless across generations) Asian female flight stewardess with gentle smile dressed in traditional batik was used to differentiate the airline's brand based on the delivery of both in-flight services and ground services. Even today, the Singapore Girl remains synonymous with SIA and its personalised quality service. The Singapore Girl is a central idea of the airline's marketing strategy. Her photographs, along with slogans such as 'A great way to fly', have been used to effectively market the airline, making the Singapore Girl, launched in 1973, one of the longest running advertisement campaigns. Most passengers recall the Singapore Girl advertisement, even if they have not flown with SIA. According to *The Straits Times* of Singapore, removing the Singapore Girl icon from the airline would be like 'removing Mickey Mouse from Disneyland'. The figure of the Singapore Girl is so strong, that it gained a position in Madame Tussaud's Museum in London, in 1993. The Singapore Girl is placed at the forefront of SIA's branding and advertising campaigns, which evolved and transitioned to messaging on fleet expansion, routes, features and in-flight services in the 2000s. It remains an important diversification factor amongst retail and business travellers and strengthens the airline's employee value proposition as well.

SIA's differentiation strategy is based on in-flight services. Onboard the airline provides food of high standards, and, most importantly, it serves it with warm smiles. SIA has an International Culinary Panel of seven chefs who designed special attractive dishes. As ingredients change according to season, they created new dishes from different places around the world for the Epicurean Gallery and introduced them into the menu in Suites, First Class and Business Class. SIA has international wine tasters to advise the airline on an array of French, American and Australian wines to be offered to passengers in each class; moreover, air sommeliers recommend a suitable wine to accompany the meal. In addition to the good food, SIA passengers are also treated with personal attention.

SIA also gives First Class and Business or Premium Economy Class (according to the type of aircraft) passengers an opportunity to have their preferred dishes on the journey, on prior notice to the airline, and serves a wide range of meals for kids from different cuisines. Kids can select their preferred meal before the flight for a special treatment. Though the airline's competitors come up with similar facilities of onboard food, the service-minded attitude with which SIA serves the passengers made it difficult for the competitors to compete against it effectively. SIA crew members always have a 'can do' attitude and they outperform to satisfy passengers. Passengers are served with small bottles of Evian water that can be kept safely by their side even when they fall asleep.

SIA has a frequent flyer programme called KrisFlyer, rewarding its members when they travel, enjoy experiences and order SIA meals at home. A member of this programme earns a specific number of miles every time the member card is used. These miles can be converted to flights and upgrades, or used to save money on hotels, car rentals and unique experiences. The KrisFlyer Programme has three membership tiers according to the level of loyalty: (a) KrisFlyer which provides advantages such as discounts on flight add-ons, onboard Wi-Fi and exclusive KrisWorld content; (b) KrisFlyer Elite Silver with advantages such as bonus on actual miles, complimentary seat selection and all the lower tier features; and (c) KrisFlyer Elite Gold with greater privileges such as lounge access, priority check-in, baggage handling and all those associated with the lower tiers.

SIA looks after the comfort of its customers onboard. The seats are designed to be comfortable and restful. Whenever a passenger in business class wants to rest, their seat can be converted into a flat bed along with a mattress, duvet and pillow, ensuring their privacy. The airline also provides sleepers and toiletry kits for a comfortable journey. Desk facilities and inflight Wi-Fi access are also provided to passengers travelling in Suites, First Class, Business Class, PPS Club members and KrisFlyer members. Passengers can also read news and other interesting topics from a varied selection of magazines and newspapers.

The airline also provides an entertainment system called Krisworld. There are individual video screens for each passenger in each class; the system includes 1,800 on-demand entertainment options such as movies, TV shows, Kids TV, music albums and interactive games. SIA launched a new entertainment system which offers passengers a choice to modify their entertainment experience by signing in as a KrisFlyer member. This application focuses on loyal passengers, who can also use the free SingaporeAir mobile app that also gives the opportunity to enjoy KrisWorld. The enhanced mobile app features an improved personalised experience onboard, providing access to the e-Library and real-time information for the trip. The use of these systems encourages the contactless communication with cabin crew. Moreover, it provides features such as KrisPlay, a loyalty digital wallet, KrisShop, translation assistant, capture and discover and baggage measurement, allowing the passenger to use and carry the SIA brand name throughout their travel. Likewise, the SIA cabin crew has access to the BEST application, to enhance customer satisfaction in premium cabins. The aim of this application is to note personal customer preference and provide customised services beyond expectations. Thus, premium passengers will find their personalised preferences onboard irrespective of the flight route and crew.

SIA believes in proactive customer engagement and service excellence, investing in features that add value. To respond to the demands of the times, KrisFlyer unveiled a new mascot in 2019. A global social media campaign to name the mascot was initiated to further reach out to millennials (i.e., those born between 1981 and 1996). The mascot was eventually named Myle and a subsequent marketing campaign was launched to remind members that miles can be earned easily through their everyday shopping. Various campaigns and promotions are organised, targeted to reward loyal passengers. SIA focuses on providing unforgettable and consistent experience to customers at every touchpoint, following a customer-oriented strategy.

SIA is not appreciated only for its excellent in-flight services, but also for the quality of its ground services. It introduced novel ideas to make ground service more efficient and useful to its customers. SIA uses every opportunity to improve its customer service on the ground, by using contactless technology. Check-in services are also very flexible as passengers can check-in two or three hours (for long-haul flights) before the departure time. Passengers can also auto check-in online or via mobile app to 48 hours before the flight's scheduled departure and use bag drop counters. Passengers using auto check-in can select their seat in advance, choose meals and/or add baggage. Moreover, the boarding pass is issued in digital or mobile format. Online services are provided to passengers with information on flight status, schedules and lost baggage tracing. SIA provides resting areas to their customers, with some of them offering five-star treatment. These areas include the first-class check-in reception, the SilverKris Lounge, the KrisFlyer Gold Lounge, Partner lounges and the SilkAir Lounges. Members of KrisFlyer Elite Silver and KrisFlyer Elite Gold are invited to visit the KrisFlyer Silver Lounge and KrisFlyer Gold Lounge, respectively, on a complimentary basis. Lounges operate on a 24/7 basis providing snacks and beverages, internet access, private entrance, shower and a comfortable and isolated environment. SIA provides premium customers with a meet-and-greet service at their hub Singapore Changi Airport, with buggy transport within the terminals.

Based on the previous analysis, Figures 7.1 and 7.2 depict two alternative SIA customer profiles. The *Xiu Zhāng* persona corresponds to a typical business traveller while the *Tara Wang* persona corresponds to a leisure passenger.

It is also worth noting that SIA launched a series of services to increase revenue and respond to new travel conditions, further to the reduction or cancellation of its customers' travel plans due to the Covid-19 pandemic. To create an innovative customer experience, the airline launched 'flights-to-nowhere' in 2020; nonetheless,

FIGURE 7.1 Singapore Airlines Persona (Source: Developed by the authors).

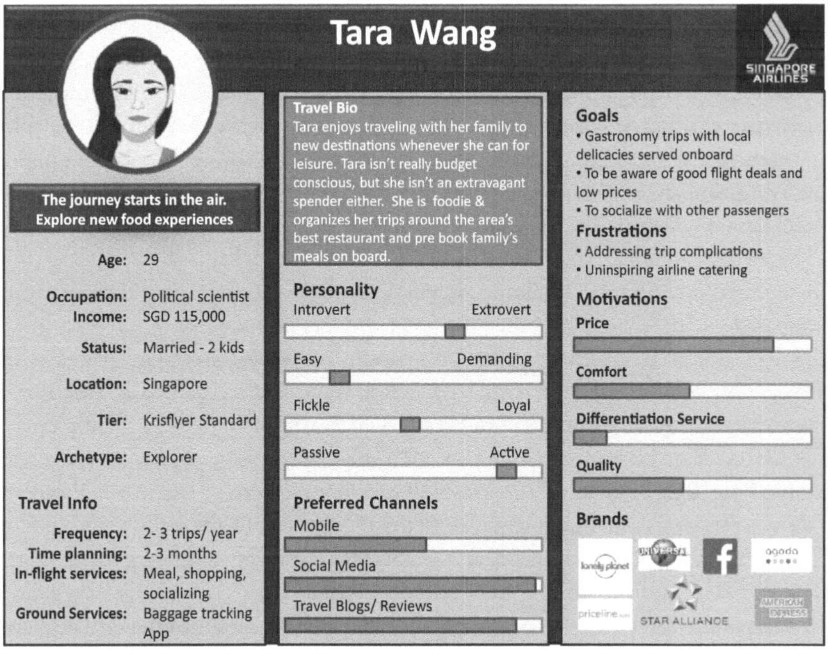

FIGURE 7.2 Singapore Airlines Persona (Source: Developed by the authors).

these were discontinued a few months later because (among others) of negative environmental considerations. In addition, an A380 aircraft has been turned into 'Restaurant A380@Changi' for a limited period, allowing passengers to enjoy a meal cooked by a famous Singaporean chef and experience a tour of the aircraft. Building on this, the airline has also launched a new service enabling customers to taste all flights' menus at their home, either delivered or by booking the chef to cook privately and serve the meal. The home-dining experience is paired with the choice of wine from SIA sommeliers and the meal includes signature starters. This experience targets customers who desire to taste and live the experience of First Class or Business Class meals at home or to recall their pre-Covid-19 flight moments. The home meal is accompanied with suitable tableware and luxurious amenity kits (Singapore Airlines website, 2021).

Behind-the-scenes tours of the airline's training facilities with a range of activities including flight simulation also proved very popular. From a B2B perspective, the airline launched the Singapore Airlines Academy in 2020, offering a range of tailor-made training programmes in service excellence, operational excellence and digital innovation to organisations and businesses across various industries. The Academy relies on in-house certified instructors and facilitators. Furthermore, SIA has started operating the Krislab, as evidence of the brand's commitment to innovation and how to apply different types of innovation.

Scoot

Scoot was launched by SIA Group in 2011, as a full subsidiary of Singapore Airlines. Network and other overlaps in their operations are rather rare. Scoot serves low-cost medium- to long-haul leisure markets within a twelve-hour radius of Singapore, where the existing Singapore Airlines and SilkAir products are either not suitable, or where the market is large enough to support both a full-service carrier and a low fare airline. The operations and key decisions made at Scoot are managed completely independently from Singapore Airlines.

Scoot is a member of Value Alliance and operates a fleet of 49 Boeing 787 Dreamliners and Airbus 320-family aircraft. These types of aircraft are more spacious compared to those used by other low-cost carriers and are comparable to those flown by full-service airlines. Most destinations served by Scoot are places with all-year-round leisure and/or business demand. Scoot was recognised as 'Best Low-Cost Airline' by AirlineRatings.com between 2015–18, and ranked in the top ten of the 'World's Best Low-Cost Airlines' in 2015 and 2018 by Skytrax. Moreover, it was awarded the 'Best Low-Cost Carrier' 2019 prize by the Travel Weekly Asia Readers.

The company's website and social media clearly set out the Scoot concepts of 'You only pay for what you need' and 'Escape the ordinary' and define their target customer. With a funky attitude – Scootitude – the carrier declares that it is an airline for the young, the young-at-heart and the value-seeking. Scoot offers its modern services with a distinct personality and identity to people who travel for discovery, connections and experience. Passengers have the choice to pay for only what matters, tailored to their specific needs. Emphasis is placed on cost reduction with a no-frills service onboard. Target customers consider a flight as a way of getting to a destination and not as the main element of their vacation experience, i.e., Scoot customers are willing to pay less for their transportation to be able to spend more money at the destination. They can just buy a seat or add anything from a meal to baggage allowance, Wi-Fi access and extra legroom. Scoot has bundled the most common options, providing a set of add-ons with discount: *Fly*, *FlyBag* and *FlyBagEat*.

Scoot is known for its unconventional and impressive campaigns and the successful use of social media to communicate with its passengers. It has gained customers from the start of its operations, creating an audience on social media. Scoot actively generates brand awareness using fun and a playful spirit. Passengers can participate in competitions, promotions and gain free seats, spreading the brand name through e-WoM, i.e., electronic Word-of-Mouth. Slang and wordplay are used to attract new fans and retain the existing ones via the cool Scootitude and viral content. A campaign 'Get Outta Here' created social media buzz. One of the latest campaigns on Facebook rewarded ten fans who were fast enough to respond to a post with complimentary in-flight meals delivered to their homes. This was a great opportunity for those who had missed flying with Scoot and wanted to live the experience of onboard dining in the comfort of their own home, recapturing their pre-Covid-19 travel experience.

In line with other low fare airlines, Scoot offers its passengers lower-than-usual fares in return for a lower level of service. The basic fare secures nothing more than a seat onboard. Nonetheless, the airline does not cut any corners with respect to safety and cabin hygiene practices, which is of utmost importance during the Covid-19 pandemic. The concept of a 'simple service model' also reflects a more general point about eliminating the 'frills' which are often costly services and characterised by complexity in their management, such as special meals, interline cooperation agreements with other airlines and cargo/freight carriage. Passengers can choose a specific class, Scoot Economy or ScootBiz. Scoot Economy can be upgraded by choosing Super or Stretch seats with adjustable headrests. Passengers of economy class can travel in the ScootinSilence quiet cabin zone (where children are not allowed) and are provided with additional privileges such as early disembarkation from the aircraft. ScootBiz targets customers who are looking for convenience and full services from meals to entertainment.

All customers can use the Scoothub, an in-flight portal, on their own device to reduce personal contact with cabin crew during the Covid-19 crisis. To encourage this practice, the airline offers amenities including onboard Wi-Fi internet connectivity and in-seat power plugs on selected flights. This makes Scoothub the one-stop in-flight shop solution for contactless experience. Passengers can enjoy diverse functionalities and order a meal, shop from duty free amenities, read Scoot magazines, enjoy the selection of games and access the route maps. They can also sync their Krisflyer membership to earn or redeem miles when they purchase items via KrisShop or activities and attractions via Pelago, the SIA Group's new platform for destination inspiration and content. This offers viewing access to organised parties with live music, acapella music, neon parties and dancing. It is remarkable that Scoot promotes such an onboard fun spirit, even though its target customers do not consider the flight as part of their travel experience.

Scoot uses advanced technology in ground services too, aiming to reduce operating costs and to provide contactless services such as check-in. Customers can book a flight using a transactional chatbox, named M.A.R.V.I.E. on Scoot's Singapore Facebook page via Facebook Messenger. Passengers receive an e-mail containing their travel details and confirmation number when they book online, so they can travel ticketless, using their mobile. This helps to significantly reduce the cost of issuing, distributing, processing and reconciling millions of tickets each year. Scoot has a frequent flyer programme called Scoot Insider, rewarding members with exclusive promotions and privileges. A member of this programme has a personalised profile and is offered birthday discounts, travel management, promotional deals and information on the airline. Moreover, Singapore Airlines Group has integrated the airlines' reward/loyalty systems. So, Scoot passengers can redeem and accrue Singapore Airlines KrisFlyer miles. Passengers who use Scoot in Style can enter Scoot Lounges in Changi Airport. Lounges operate on a 24/7 basis and provide self-service buffet and beverages, internet access, showers and a comfortable environment. Further to the Covid-19 outbreak and to meet the safety commitments, the lounges' self-service buffet meals have been suspended (Scoot Airline website, 2021).

Based on the previous analysis, Figures 7.3 and 7.4 depict two alternative Scoot customer profiles. The *Pat Whelen* persona corresponds to a typical budget-conscious traveller, while the *Irfan Sayid* persona corresponds to an explorer.

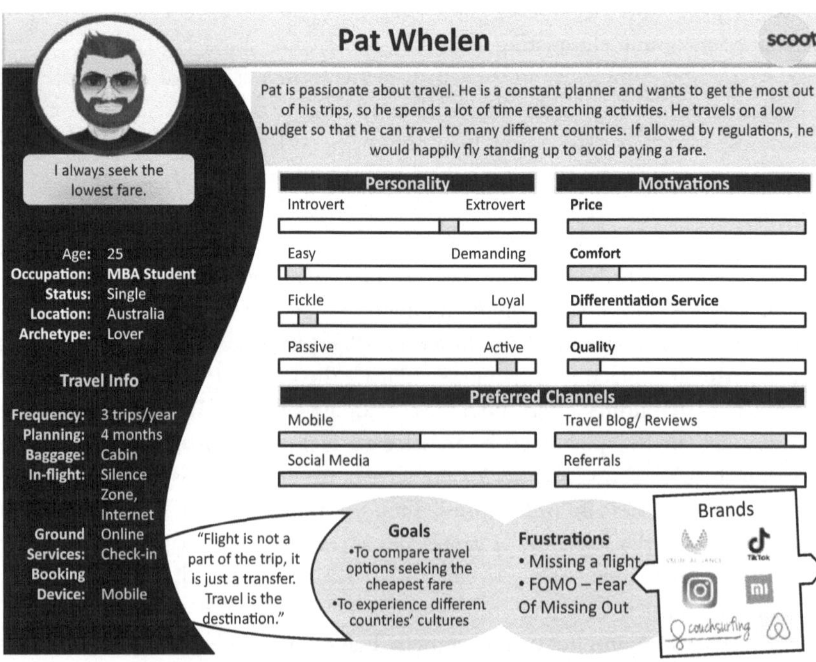

FIGURE 7.3 Scoot Customer Persona (Source: Developed by the authors).

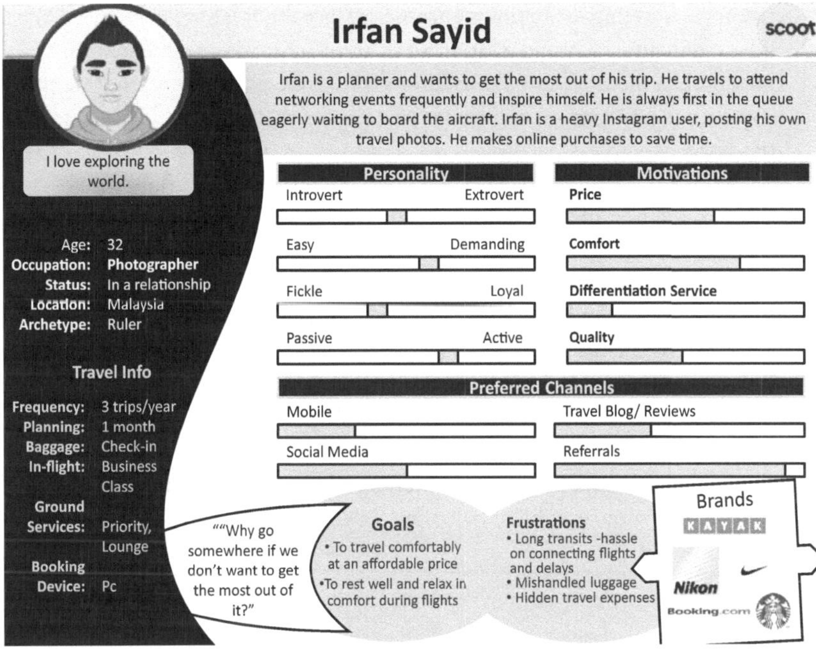

FIGURE 7.4 Scoot Customer Persona (Source: Developed by the authors).

Conclusion

This case study focuses in detail on the differentiated services and targets of the SIA Group to develop a deeper understanding of the strategic functioning of a heavily diversified airline within an airline. Marketing strategies are essential for airlines to ensure survival and profitability, and airline managers should develop related skills to successfully address the challenges posed by the Covid-19 pandemic. The SIA Group is an interesting example of a complex organisation in the air transport industry and is also a proper illustration of the role of the Segmentation, Targeting and Positioning (STP) model in tourism.

The SIA Group attracts international visitors from different market segments. The marketing messages are clearly designed to appeal to these distinct markets. Position strategies are important because they determine how customers will view the product or service offered and create a special position for it in their mind. The SIA Group's positioning and commitment to the brand remains strong post Covid-19. The challenge is to keep delivering on the high-cost promise of quality, innovation and service, providing stand-out services.

Questions and self-assessment

1. Identify the various airline market segments and discuss the segmentation criteria used by the SIA group. What are the criteria/ benefits sought by the

customers traveling for business purposes? What about leisure customers? What is the competitive advantage of Scoot in terms of market segmentation?

2. Considering the new travel conditions set by Covid-19, consumer demands have changed. Design one customer persona for each airline and evaluate their attractiveness.
3. Discuss the targeting strategies followed by Singapore Airlines and Scoot. Which is the best in your opinion? Justify your answer.
4. What processes, criteria and methods of market segmentation and targeting are used by the airlines of the SIA Group?
5. What is the Unique Selling Proposition (USP) of each SIA Group airline for the market segments it serves?
6. What are the comparative advantages of the SIA Group airlines, and how are they positioned to address the various customer segments? Create a perceptual map based on two key characteristics.
7. Discuss the SIA Group airlines positioning strategy. Can you identify potential gaps that have not been filled by competitors and may be considered as potential opportunities?

References

Scoot Airline website. (2021). Flyscoot.com/en

Singapore Airlines website. (2021a). www.singapore.com

Singapore Airlines website. (2021b). Annual Report FY 2019/20. www.singapore.com

Further reading

Fyall, A., Legohérel, P., Frochot, I., & Wang, Y. (2019). *Marketing for Tourism and Hospitality: Collaboration, Technology and Experiences*. New York: Routledge.

Kotler, P., Bowen, J. T., Makens, J., &Baloglu, S. (2017). *Marketing for Hospitality and Tourism*. India: Pearson.

Case 8

A UNIQUE TRAIL

Is the product ready-made?

Małgorzata Zdon-Korzeniowska and
Monika Borowiec-Gabryś

Duration

The time needed to complete the task is approximately 135 minutes, with the following distribution of tasks over time:

- Introduction to the task (instructor) – 10 minutes.
- Reading and group discussion of the case study text (including answers to questions from all four parts) – approximately 50 minutes.
- Instructor-moderated discussion – 20–25 minutes for parts I and II and about 10–15 minutes for parts III and IV.

Learning objectives

Upon completing the case, participants will be able to:

- identify the stages of tourist product development in realtion to the concept of travel routes
- critically assess the consistency and validity of the marketing tools used with a defined tourist product
- adopt the tourist's point of view in assessing marketing activities taken in connection with the preparation and commercialisation of a tourist product
- engage actively in teamwork and be open to discussion
- analyse sources of successes and failures of tourist products.

Target audience

This topic, with some modifications, may be introduced to the audience to advance their marketing knowledge and skills as well as their knowledge of the

DOI: 10.4324/9781003182856-8

socio-economic conditions of regional development. Students should have basic knowledge of the field of tourism marketing and socio-economic geography, as well as some basic group work skills. This case best suits tourism and recreation-related courses in higher education institutes, during which students learn about the principles of tourist product development.

Teaching methods and equipment

Group work, moderated discussion, brainstorming, snowball method, use of computers or smartphones, website analysis and evaluation.

Teaching instructions

At the outset, the instructor introduces students to the WA Route's tourist product and the concept and plan of the classes. Next, students are divided into groups (max. three to four persons), where, after reading the case study in a group, they answer the questions for discussion (using activating methods such as brainstorming or a snowball method). The questions have been divided into four themed parts. After tasks in each part are complete, a discussion moderated by the instructor follows, leading to selecting agreed summarising answers/solutions to the questions asked. As such, the anticipated outcomes of the discussions should include the following:

- *Part I*: stages of creating the WA Route and an attempt to sketch a universal process of shaping this type of tourist product.
- *Part II*: a list of proposals for other (not included in the described case) marketing activities (e.g., following the 7P concept).
- *Part III*: proposals for changes or necessary information, which should be added to the website informing about the WA Route and ideas in broader use of digital marketing tools.
- *Part IV*: a list of features of a regional tourist product and examples of this product type.

Case

Nowadays, we can observe the growing importance of tourism in accelerating socio-economic development and levelling the disproportions of different scales of spatial systems. With increasing frequency, tourism is an essential factor in developing tourist-attractive localities or regions and an effective tool for activating areas where the tourism function has not played an important role so far or has not existed at all. Activities involving servicing tourist traffic are becoming an alternative to unprofitable agricultural activities, unprofitable health resorts or industrial agglomerations. Tourism is an alternative to other development trajectories; the selection and application in a given area would be unfavourable, or impossible, for various reasons. Apart from stimulating the socio-economic development of towns or regions, tourism also contributes to discovering and cultivating cultural and natural values and resources. Well-managed tourism can even become a way to preserve and protect the natural, cultural and historical heritage of specific places or regions (Zdon-Korzeniowska & Noviello 2019). On the one hand, heritage elements become attractions around which unique tourist products are created. On the other hand, they become a distinctive feature of a place or region, based on which local communities build their identity and sense of belonging.

For several years in the European Union countries, a strong trend to emphasise their own regional affiliation and identify with the so-called 'small homelands' has emerged. Mechanisms of local and regional development are built to simultaneously enable the promotion and protection of particular areas' cultural and natural heritage. The concept of creating regional tourism products integrating these two factors, i.e., tourism and heritage, is relatively new – both in Polish economic practice and in literature. Nevertheless, the concept of a 'regional tourist product' is increasingly used in programmes or development strategies – both nationwide and for individual regions, which are being developed. It also appears, although much less frequently, in scientific studies and texts. However, there is no coherent, holistic approach to the concept of shaping regional tourism products. And above all, there is no clear and complete definition of what is defined as a 'regional tourist product'.

In the light of the presented premises, this case deals with the issue of the Wooden Architecture Route (the official name of the trail) in Poland, whose concepts generally respond to the broad view for defining a tourist product that recognises it as a spatial product, integrating the activities of its many co-creating entities, built on the basis of regional attractions, i.e. sites of wooden architecture, testifying not only to the construction history of the region but also to the fate of the inhabitants of these lands. These attractions form a significant part of the historical and cultural heritage of the region.

The Wooden Architecture Route (WA Route) is one of the most significant projects of setting out tourist routes in Poland due to its territorial range, subjective scale and planned time horizon. The route currently runs through four regions in the southern part of Poland (i.e., Lesser Poland, Subcarpathian, Silesian and Świętokrzyskie Regions). Note that this is the only initiative of such a large size in Europe regarding the diversity and multiculturalism contained in individual parts of the trail. It covers a broad spectrum of art and wooden architecture. The sites included in the WA Route include churches, Eastern Orthodox churches, open-air museums, manors, entire urban layouts (such as small-town or spa complexes), chapels, roadside shrines, bell towers, houses, country cottages and noble manors, wooden farm buildings (such as granaries or water mills), inns, foresters' lodges, palaces and open-air museums. Some of them are unique 'pearls' of architecture – 14 of them have been inscribed on the UNESCO World Cultural and Natural Heritage List. Pictures of sample objects are presented in Figures 8.1–8.4.

The above-mentioned sites on the UNESCO List were entered into the List primarily because of the authenticity of the sites: the authenticity of architecture,

FIGURE 8.1 Wooden Architecture Route – St Bartholomew's Parish Church in Kraków-Mogiła

Source: Own archive

FIGURE 8.2 Wooden Architecture Route – one of the buildings in the Orawa Ethnographic Park in Zubrzyca Górna

Source: Own archive

material and construction, and the authenticity of the functions, content and rituals, as well as the conservation works they have undergone.

The WA Route was designed as a car trail, intended mainly for motorised tourists. It is currently approximately 4,262 km long and connects 532 historic wooden buildings of exceptional cultural value, located in various municipalities (Table 8.1). In each region; the WA Route is divided into 24 smaller sections, 'routes' or 'loops', that have been sectioned off.

FIGURE 8.3 Wooden Architecture Route – The Lanckorona Urban Layout and Buildings

Source: Own archive

FIGURE 8.4 Wooden Architecture Route – The Wooden Gothic Church of All Saints in Blizne (UNESCO World Heritage List)

Source: Own archive

TABLE 8.1 The Wooden Architecture Route – General Information (Quantity/Volume)

Region	*No. of items*	*Route length (km)*	*No. of sections*
Lesser Poland	253	1500	4
Subcarpathian	127	1202	9
Silesia	93	1060	6
Świętokrzyskie	59	500+	5
Total	**532**	**4262+**	**24**

Source: Own study based on: www.drewniana.malopolska.pl; http://sad.podkarpackie.travel; https://sad.slaskie.travel/; https://pl.wikipedia.org/wiki/Szlak_Architektury_Drewnianej.

The primary goal of creating the WA Route was to protect sacred wooden architecture, and only later did the idea of transforming the Route into a tourist product emerge. The main idea behind the creation of the WA Route was, therefore, the protection and preservation of the cultural and historical heritage of Lesser Poland – popularisation of wooden architecture, unique on a national and European scale, but also the world – and with a further goal to build a tourist product based on the WA Route. The delineation and marking of the trail were to encourage the local community and the tourist industry to develop it further and create a complete tourist offering. The very idea was 'born' at the end of the 1990s in the Krakow Tourism Development Agency (KART). Initially, it concerned the consolidation of approximately 40 wooden sites into a tourist trail.

From the onset, the originators of the Route assumed that it would take the form of a complete, integrated tourist product, cultivating the heritage of wooden construction. In 1999, the idea of the Wooden Architecture Route was presented by one of the originators and first animators of the Route, Edward Turkiewicz – the then president of the Krakow Tourism Development Agency:

> The trail will provide an opportunity to get to know (…) an unusual mosaic of varieties of wooden architecture, it will show its beauty and uniqueness. (…) The proposed route will not be just a sightseeing proposal. Hotels and boarding houses, tourist hostels, bars and restaurants should come to life around the Trail; handicraft in its natural form and surroundings may be reintroduced. A similar situation will arise around all tourist attractions and areas for practising various forms of active tourism: horse riding, skiing, bicycles, etc. (…) Proposals will be addressed to various entities interested in participating in this project. The project will be successful if everyone agrees that the development of tourism in this area should be supported or that tourism is a source of income, they will co-finance individual stages.
>
> *(Turkiewicz, 1999, p. 238)*

Currently, the WA Route is a supra-regional project. Apart from the Lesser Poland Region, the initiative to create the Route included three other regions: Subcarpathian Region, Silesian Region and Świętokrzyski Region. The

WA Route is also connected with a similar project in the Prešov Region (Kraj Prešovský) in Slovakia.

The process of shaping a tourist product

To illustrate the process of shaping the WA Route, including analysing its structure and activities undertaken by entities co-creating this product, only the Lesser Poland part of the WA Route was subjected to detailed analysis. It is this part that boasts the most extended history of operation. Hence the initiative to create a tourist product using historical sites of wooden architecture as attractions. Moreover, the Lesser Poland Region formally coordinates the Route and is the most experienced in developing the tourist product.

As previously mentioned, the WA Route was intended to be a complete, integrated tourist product. The guidelines of the Route were to be, apart from making available and popularising monuments of wooden construction, an axis for the development of other elements of the offer, i.e., accommodation, catering, recreation and entertainment facilities, etc. Although after about a dozen years of operation, the WA Route is still not a tourist product in the broad and full sense of the word. The Wooden Architecture Route is one of the flagship initiatives of the Lesser Poland Region in the field of shaping tourist products. Creating products using the region's cultural heritage has been included in the Development Strategy of the Region as one of the ways to support the development of the 'leisure industry'[1] – referred to as the 'industry of regional opportunity' for the Lesser Poland Region. The above lines of action are also upheld in the current planning perspectives.

At present, the core offering of the WA Route tourist product includes:

- regional attractions in the form of historic buildings of wooden architecture
- technical information infrastructure in the form of information boards placed next to the site, connection boards and unique road signs supporting the organisation of traffic on the Route.

The first stage in the process of creating the WA Route concerned design. It consisted of developing the concept and marking the route based on the identified wooden architecture sites and the existing road infrastructure in the region. The premise for creating the WA Route was to emphasise the region's existing, natural and cultural values and thus create potential opportunities to supplement the selected sections of the routes with additional attractions characteristic of a given subregion. In addition, as mentioned above, when developing the WA Route concept, it was assumed that it would become the axis for the development of the complete product offering, i.e., it would be supplemented with accommodation and catering facilities, as well as other additional attractions.

After the concept of the WA Route had been developed, and its course agreed with local governments, the Lesser Poland Region, in cooperation with the Academy of Fine Arts in Krakow, invited artists to present their concept of a graphic

SZLAK
ARCHITEKTURY
DREWNIANEJ
W MAŁOPOLSCE

FIGURE 8.5 Official Logo of the Wooden Architecture Route

Source: www.drewniana.malopolska.pl/

symbol for the Wooden Architecture Route in an open competition. The winner of the competition became the official logo of the WA Route (Figure 8.5).

The WA Route offer is supplemented by a published list containing practical information about selected accommodation and catering facilities and tourist information points located in the towns through which the Trail runs, which serves as an information guide on the Lesser Poland WA Route. The sites included in the list do not 'consciously' create the product offer of the WA Route. This list was prepared for the needs of the above-mentioned publication, based on the existing databases of accommodation, catering facilities and tourist information points of the Lesser Poland Region. These sites were 'arbitrarily' included in the offer and only as a compilation supplementing the information contained in the guide on the WA Route. No cooperation was established with the aforementioned sites in the field of co-creating the offer of the WA Route. Nevertheless, the list provides detailed information on:

- tourist information points (giving the name, address, telephone and possibly e-mail address or website)
- accommodation facilities broken down into types: hotels, motels/inns, guesthouses, excursion houses, youth hostels, mountain hostels, student huts, campsites (with the name, object category, address, telephone and e-mail address or website)
- catering outlets (giving the name, address, telephone and possibly e-mail address or website).

However, for many tourists, a certain barrier to the availability of this exclusive and comprehensive guide to the WA Route may be the relatively high price of the publication.

The complete offering of the WA Route tourist product also consists of the following elements:

- transport infrastructure: access roads to the sites and parking lots
- recreational infrastructure
- additional regional attractions other than wooden architecture
- other services and facilities.

It also consists of intangible elements/components such as the ambience and the attitude of the local population to tourism and tourists.

Initially, the Marshal's Office of the Lesser Poland Region was the operator of the WA Route in the Lesser Poland Region. Since 2008, on behalf of the Region, the Trail has been operated by the Tourist Organisation of the Lesser Poland Region as 'the Open Wooden Architecture Route'.

In the first stage of the project, apart from the logo, the WA Route concept was developed based on the identified wooden architecture items and the road infrastructure existing in the region. When developing the WA Route concept, it was also assumed that it would become the axis for the development of the full product offering, i.e., it would be supplemented with accommodation and catering facilities and other additional attractions.

Stage two involved putting up information boards and signs connecting parts/sections of the Route (see Figure 8.6 and 8.7). To this end, approvals were sought from owners of the sites to have the boards and signs installed. The content of the descriptions put up next to monuments, as already mentioned above, was prepared by the Regional Centre for Research and Documentation of Monuments in Krakow.

The next step involved the organisation of traffic on the WA Route and the appropriate traffic marking and signage of the Route. For this purpose, approximately 606 road information signs (showing directions) were designed with the WA Route logo (see Figure 8.8). Road marking was carried out on the basis of relevant regulations and agreements. The WA Route was planned as an automobile trail from the very beginning.

After all necessary on-site inspections and arrangements had been completed and the WA Route had been marked out, the Route promotion campaign began. Some advertising materials were developed in the form of folders, leaflets, maps, multimedia presentations, promotional publications, etc. – in several language versions. A website was also developed with information on the WA Route, detailed descriptions and photos of the sites.

The initial promotional activities included information campaigns about the WA Route organised in selected historic buildings – mainly in open-air museums. It was no coincidence that open-air museums were chosen as the campaign sites. Open-air museums are visited more often than other sites located on the WA Route. Moreover, they had specific 'infrastructure' related to the reception and service of tourists and visitors, such as appropriately prepared exhibitions, organisation of tours or qualified and experienced staff. Other sites included in the Trail, such as historic Orthodox churches or Catholic churches, had no experience handling tourism. The open-air museums were to act, in a way, as specific information and promotion points for the WA Route.

Another critical undertaking in disseminating information and promoting the WA Route was the opening, in collaboration with the Metropolitan Curia in Krakow, of the Wooden Architecture Route Information Desk at the Diocesan

FIGURE 8.6 An Information Board of the WA Route

Source: Own archive

Information Centre for Promotion and Tourism Office at 12, Wiślna Street in Krakow. The initiative is more prominent than just launching an information point about the Trail. Commencing cooperation with the Curia attempted to reconcile the church authorities, which operate a large part of sacred wooden buildings in the Lesser Poland Region. It paved the way for better communication and collaboration in the WA Route project. Indeed, one of the ideas in protecting and preserving wooden sacred sites is to restore them to their original functions – those related to religious worship or rituals. Since 2008, the issue of accessibility of some sites has been handled as part of the 'Open Wooden Architecture Route' project. Since then, selected sites on the Route are regularly opened to visitors in the

FIGURE 8.7 A Sign/Board Connecting Parts/Sections of the WA Route

Source: Own archive

FIGURE 8.8 Road Information Signs on the WA Route

Source: Own archive

period from May to September. Additionally, each of them employs a guide who is responsible for servicing visitors. Thanks to this, tourists coming to WA Route have the opportunity not only to see the sites from the outside, but also to admire the interior of temples and shrines. Additionally, during the summer season, some churches organise concerts as part of the 'Music Enchanted in Wood' series[2].

The last stage identified in the process of creating the WA Route tourist product encompasses product development activity. Actions planned for this stage involved:

- improving and complementing signage and marking of the Route
- re-surveying the condition of the sites – there are plans to extend the WA Route by adding new sites or exclude those that, for example, have suffered considerable devastation and harm the image of the Route
- creating a positive image and promoting the brand of the WA Route
- introducing additional activities promoting and triggering visits to the Route, such as night tours of the sites, open days of the WA Route. To increase the attractiveness of the offer, cyclical events in selected places and sites of the WA Route are planned, such as weekly concerts in the summer season, from mid-June to mid-September. The first such pilot entitled 'Folk Chopin – at the sources of Frederick Chopin's music' was organised in 2006 and included a series of concerts presenting Polish folk songs and melodies that could have inspired F. Chopin in his compositions. Polish dances such as *oberek, kujawiak, polka* or *mazurkas,* arranged and played on authentic folk instruments, enjoyed great popularity, attracting visitors to wooden architecture (e.g., in Łopuszna or Łomianki).

Analysing the activities to date in connection with shaping the tourist product

For the WA Route, by far, promotion and image building activities were the marketing instruments employed in the promotion. There is indeed no developed and written strategy in this regard, and the actions taken resulted from the general guidelines adopted in the Regional Strategy. However, the WA Route was promoted by employing a broad range of forms and tools, addressed to both internal and external customers, including:

- advertising in several language versions, in the form of:
  - guidebooks and information leaflets with a description of both wooden structures and the Route
  - a tourist map of 'Lesser Poland - the Wooden Architecture Route' at scale 1:200 000
  - a multimedia presentation of the WA Route on a CD-ROM
  - a photo album entitled 'The Wooden Architecture of Lesser Poland'
  - the 'Wooden Architecture Route' tourist guide containing a detailed description of the trail and wooden architecture sites, accommodation and catering facilities, tourist information points and information on cyclical events
  - promotional events, including concerts or open days, organised in wooden churches
- a website with a description of the sites and a map of the WA Route at www.drewniana.malopolska.pl (read on 08.03.2021), available also in several language versions

- public relations campaigns, which include the opening of a tourist information point about the WA Route in the Diocesan Information Centre for Promotion and Tourism, which potentially paved the way to better cooperation with parish priests – the owners of a significant part of wooden buildings
- other forms of promotion, for example cooperation with partner regions including Subcarpathian, Silesia and Świętokrzyskie Regions.

Moreover, to distinguish the offer of the WA Route and give it a clear, positive image, a consistent system of visual identification of the WA Route is used. The trail has its logo and a uniform marking system for trails and wooden sites, making it easier to recognise the Route's offering and place it in the minds of recipients.

For the activities involved in developing the offering of the WA Route to be effective, a marketing strategy, or at least a marketing plan, should be formulated, taking into account a properly composed set of marketing impact instruments. However, the success is dependent on their coherent application without giving preference to one instrument at the expense of another.

After more than two decades of operation, the planned target development of an integrated tourist product has not been achieved. Many tourists, particularly foreign ones, encouraged by the promotion, did not have the opportunity to try the WA Route's offering, probably because of its poor accessibility (transportation/ opening hours). Tourists interested in the WA Route receive professionally published folders and guides of the Route and wooden architecture sites with numerous photographs. Still, without their own means of transport, they are unable to try and use the offering.

Foreign tourists may be discouraged by the limited transport accessibility of the Route both in terms of transportation and easy access to sites. Even if a foreign tourist travels by car and the WA Route is well marked, still it could be a relatively big challenge to visit monuments of wooden architecture scattered around the region with a distance of a dozen or several dozen kilometres, in particular when the tourist is not familiar with the region and doesn't speak Polish. In addition, some churches are not easy to visit because of restricted access – in general, they are not oriented towards servicing tourist traffic and do not have, for example, visiting hours, which in some cases requires calling ahead, identifying the parish priest and arranging a date to visit to the site. Language is another limitation in accessing the churches, in particular in the case of foreign tourists.

Questions for discussion

Part I: The process of creating a tourist product

1. Identify a tourist product creating/development process. Identify and briefly characterise stages of this process.
2. What are groups of entities engaged in the process of creating the comprehensive tourist product of the WA Route? Indicate actions/solutions to potentially

improve the coordination of the complex offering of the spatial tourist product of the WA Route.

Part II: Potential and used marketing tools

3. What marketing tools have been used? What is your opinion on their coherence, adequacy and balance in the above-described case in the context of the 4P formula?
4. What actions should be taken to improve the accessibility of the WA Route, which, before anything else, is the precondition of its effective commercialisation?
5. What other marketing tools/campaigns (for example based on the 7P concept) could be suggested? Use brainstorming or a snowball method to create a list of such actions/tools.

Part III: A critical assessment of the website

6. Taking the role of a tourist interested in visiting the WA Route, plan your trip using the information provided on the website www.drewniana.malopolska.pl. How useful is the information contained therein for a potential tourist interested in visiting the WA Route?
7. What other data/information should be added to the website to make it more useful to a client-tourist?

Part IV: Characteristics and examples of regional tourist products

8. With reference to the Wooden Architecture Route, define a regional tourist product. List its features.
9. Give examples of other regional tourist products you are aware of.

Notes

1 *2000–2006 Regional Development Strategy of Małopolska Region*, Marshal's Office of the Małopolska Region, Kraków 2000, [e-document], www.wrotamalopolski.pl, p. 58. This objective is upheld in the Strategies for the coming years.

2 To learn more about the festival, see: www.drewniana.malopolska.pl/?page=muzyka&run=multilang.

References

Turkiewicz, E. (1999). Szlak Architektury Drewnianej [Wooden Architecture Road]. In: *Turystyka w obiektach zabytkowych i sakralnych [Tourism in historic and sacred buildings]* (p. 238). Kraków: Górnośląska Wyższa Szkoła Handlowa [Katowice Business University], Instytut Turystyki [Tourism Institute].

Zdon-Korzeniowska, M., & Noviello, M. (2019). The Wooden Architecture Route as an example of a regional tourism product in Poland. *Sustainability*, 11(18). https://doi.org/10.3390/su11185128

Further reading

Cirikoviý, E. (2014). Marketing mix in tourism. *Academic Journal of Interdisciplinary Studies,* 3(2), 111–115. Doi:10.5901/ajis.2014.v3n2p111

Čuka, P., Dorocki, S., & Rachwał, T. (2015). Development of tourism infrastructure in the regions of Central and Eastern Europe. Conference: Innovation Vision 2020: from Regional Development Sustainability to Global Economic Growth, 25, 402–417. www.researchgate.net/publication/277018032_Development_of_tourism_infrastructure_in_the_regions_of_Central_and_Eastern_Europe

Nizioł, A., & Życzyński, N. (2020). An increase of the Region's competitiveness through effective tourism product management: An example using the thematic trail. *Humanities and Social Sciences*, 27, 41–48.

Pearce, D. G. (1988). Tourism and regional development in the European community. *Tourism Management*, 9(1), 13–22. https://doi.org/10.1016/0261-5177(88)90054-4

Case 9

KEEPING UP WITH IMAGE MARKETING DURING A PANDEMIC

Monika Barnwal and Vijay Kumar

Duration

Between 90–120 minutes.

Learning objectives

Upon completing the case, participants will be able to:

- explain the importance of being proactive in image marketing during a pandemic
- identify important stages of conducting successful webinar meetings
- develop skills in conducting webinars and act as a spokesperson.

Target audience

The case aims to discuss image marketing through an effective and organised webinar meeting in a time of crisis like the Covid-19 pandemic, but could be applied to many other tourism related issues.This will be helpful for the participants in developing their managerial and marketing skills. It will also help them gai confidence in facing online webinar meetings which have gained momentum during the pandemic. In different educational programmes and higher education, the case will help in providing basic information and knowledge on marketing strategies, tourism management, destination branding, image marketing and social networking. This will help students to have a better judgement of the market and be quick to present with alternative solutions suitable for the ongoing crisis. The case starts with a video of the airline's image marketing campaign ,which is followed by a lecture on the importance of image marketing strategies,

DOI: 10.4324/9781003182856-9

as well as the need to be proactive in building relationships with Influencers and maintaining them through using different social networking channels. The application of the case also follows the importance of webinar meetings during the Covid-19 pandemic. Instructors are advised to discuss the key points and instructions to be followed while conducting online webinar meetings to avoid mistakes made by participants.

The case is designed for a group of 10–15 participants, and participants need to be divided into two teams; the Airline team and the Influencer team. The Influencers are the Social Travel Influencers who have a huge number of people following their travel expeditions on their social media accounts. In case there is a larger group, different image marketing scenarios of different travel brand settings can be discussed. The participants are allowed to choose their respective team. For a larger group, the first participating group can switch to the other team and are required to be informed and prepared for both, as an Airline team for their image crisis scenario, and as an Influencer team for their doubts and queries.

Teaching methods and equipment

The case study uses teaching methods conducted through scenario-analysis, class participation, role play and group discussion. The case is based on Social Travel Influencers who are interested in Thai Airways' new image marketing campaign, which is awarding miles points to its members by staying at their home for every four hours and is marketed through different social networking channels. So, for an effective execution of the scenario, the group would be seated with the two teams facing each other. Equipment like a projector and laptop/computer/smartphone are needed to show the video of the image marketing campaign ad in order to initiate the discussion by the Airline team.

The case includes two webinar meetings to be held. It is hoped that this initiative will make students more capable and confident in participating in online meetings. For the webinars to be held, students need to either be in a computer laboratory or they can use their laptops or smartphones to access online meeting platforms like Google Meet or Zoom. This exercise is needed to make students aware of the change the pandemic has brought into our learning environment as well as the importance of innovative image marketing during a difficult time.

Teaching instructions

To execute the case, teaching instructions are given in detail which need to be followed by the group. The instructions can be changed as per the number of participants. In case of a larger number of participants, more teams will need to be created. As the case starts, instructors should give an introduction about the importance of image marketing through social networking platforms to keep the interest among its Influencers. Instructors may mention that along with image marketing campaigns, different steps, such as adopting sustainability, etc. can also be discussed.

Step 1: Initiation

There will be two teams in each group, consisting of ten participants who are randomly selected to participate in the discussion. The Airline team and Influencer team will consist of five participants each and need to be familiar with the proposed case scenario. The Airline team is asked to start the discussion by explaining their image marketing strategy and their campaign. This team would need to decide on a spokesperson for the airline and its marketing associates. The participant selected for the spokesperson of the airline is expected to have leadership qualities, as well as knowledge of the current situation. The team needs to be clear as to their marketing agenda and its terms and conditions. The team also need to take every situation into consideration in order to be able to respond to the Influencer team's queries properly.

All participants of Influencer team will play the roles of members (already registered with airline) and non-members (not registered with airline) of the airline, from different countries. The Influencer team needs to assign the roles of airline registered Influencer and non-registered Influencers among the participants. They need to prepare questions and queries according to their roles. The aim of the Influencer team is to raise various doubts regarding the new campaigns reversing the traditional miles points programme. The time allocated to the teams is around 15 minutes to get them prepared for the webinar meeting.

Step 2: Conflict/Power battle

Proceeding to Step 2, the online webinar starts. The spokesperson of the airline along with his/her marketing associates takes their respected place in the class. The instructor should introduce the case by explaining the importance of image marketing, how the meeting will proceed and introducing the spokesperson of the airline. The spokesperson should start the discussion by showcasing their recently launched Stay Home Earn Miles points programme which is a reversal of the traditional miles points allocation, calculated by air travel frequencies of its members. They will give a brief explanation of how to get registered and utilise/redeem these points. After the explanation, the instructor should start the question/answer session. The participants in the Influencer team sould introduce themselves. After question/answer session, the instructor should end the webinar. The time taken would be around 25–30 minutes.

During Step 2, participants from both teams need to take individual notes on the points shared and discussed, as well as how other participants have played their roles. They need to note down pros and cons of the meeting along with behavioural attitudes of the participants, especially the spokesperson of the airline. At this stage, some mistakes will be noted and further evaluation of the meeting will be needed.

The first webinar meeting needs to cover three objectives:

1. *Creative thinking*: The airline has pro-actively collaborated with a marketing agency to create marketing campaigns that are out of the box and are creative

to attract the attention of its active and potential customers. The airline has adopted a responsible attitude towards their members.

2. *Image marketing campaign*: The airline is the only company to contact if any conclusion and doubts arises with their customer feedback form available on their web portal. The next webinar meeting would be scheduled with the result of the campaign, and many other campaigns relating to travel safety of customers will be introduced.
3. *Understanding*: The spokesperson needs to show understanding of the downfall of the tourism industry as well as travellers who needed to stay home despite wanting to travel. The airline needs to assure the travellers that by participating in such campaigns, they would be able to invest the points in their future travel, making their travel easier and budget-friendly.

While following the above described objectives in the discussion, some mistakes are likely to be made by participants:

- Stating incorrect facts and statistics while expressing their opinions.
- Showing aggressive and rude behaviour, especially by the Airline team, when not able to clarify the Influencer team's doubts and queries.
- Participants may get nervous or feel shy in expressing their opinions.
- The Airline team might not able to convey their message as a responsible, caring and trustworthy airline; this may result in disinterest among the Influencer team.
- The participants may not seem serious about the case and carelessly state their opinions

Step 3: Collaboration

After the webinar ends, the teams break out and discuss the case study, points made during the webinar and evaluate their notes with their fellow team members. The evaluation is done on the basis of each team's performance, their behaviour during the discussion and how much each participant was involved in sharing their opinions. Moreover, the Airline team is given new information on marketing developments and are asked to prepare for another webinar. The Influencer's team would not be provided with any information but should be told to be ready to attend the next webinar along with their new questions as per the outcome of the first webinar session. In the second webinar meeting, the role of Airline spokesperson can remain with the same participant, or, if decided by the group, can be passed to another participant. This step should last around 15–20 minutes.

Step 4: Execution/Performance

Step 4 will start with a second webinar meeting. The wbinar should start with a small introduction from the instructor and a brief summary of the first webinar meeting by the airline's spokesperson. They should point out that the Airline team will be sharing information regarding their campaigns and other programmes with the influencers. The webinar should then proceed towards a question and answer session, after which is should be ended by the instructor. The allocated time for the second meeting would be 25–30 minutes.

At this stage, key points should be noted:

- For the Second webinar meeting, the scheduled date will be on 11th July 2020 so that the result of the campaign can be discussed and suggestions can be implemented in future upcoming programmes.
- In order to avoid more confusion, the spokesperson of the airline should provide the customer support number and email address to facilitate contact with the airline over queries and feedback.
- The Airline team need to be updated about the current tourism situation and need be ready for any kind of doubts and queries raised by Influencers.

Step 5: Conclusion

Step 5 concludes the group discussion of the given case study. In conclusion, an evaluation of participants on an individual basis should be carried out on their participation and involvement in the discussion. Each point that was asked or discussed, whether positive or negative, should be analysed. The instructor can then evaluate and conclude the discussion by summarising his/her assessment and suggesting additional cases and study materials related to the discussed case study in order to gain a clearer view of the scenario. The discussion should take around 15 minutes.

The instructor should point out some important points regarding the group discussion:

- Airlines need to come up with better solutions in order to increase air travel and recover their losses.
- Airlines need to build trust with their Influencers and members, irrespective of the pandemic, and provide their members with more discounts and lucrative offers to encourage them to travel more.
- Organisations should focus on developing ongoing, consistent and trust-based relations with the media.
- Launching of image marketing campaigns in order to attract their customer's attention should not be limited to during a pandemic.

Case

First part for the Airline team

Thai Airways is the national air carrier of Thailand, a popular Southeast Asian country. The airline is well-known for its hospitality, catering service, cooperative staff and hygienic cabin. It has been awarded the most hygienic in-cabin management system by the World Health Organization, which was first of its kind in 2004.

Royal Orchid Plus is Thai Airways' membership programme, where two types of miles points are awarded to its members. The miles are earned on the basis of the paid class and flying frequency of its members. The programme has more than two million registered members.

In 2019, the Thailand aviation industry, which is mostly comprised of Thai Airways, witnessed 165 million air passengers with a 1.5% increment on the previous year. In 2020, the airline suffered heavy losses due to the global Covid-19 pandemic, which resulted in total lockdown of the country. The tourism industry suffered a massive downfall in 2020 as did the aviation industry. Amidst the losses the airline faced, the airline team kept working towards solutions to retain its image among its members as well as new potential customers. The airline, with the collaboration of Wunderman Thompson, launched a creative campaign on 24th April 2020 – the Stay Home Miles Exchange programme and posted it on various social media platforms. The theme of the campaign is the exact reversal of the traditional miles point programme in which members' gain miles points depending on the class they fly and frequency of their air travel. The campaign focussed on requesting that people stay at home and be rewarded miles points every four hours.

Geo-location technology is used to track the positions of their members. This programme was exclusively for registered Thai Airways Royal Orchid Plus members. In addition, new members could be added by simply registering on their website. The members needed to install a mobile application, and automatically their movement within a 100-metre radius was tracked. The campaign lasted until 23rd May and collected miles would then be transferred to members' Royal Orchid Plus account by 10th July 2020. The app also rewarded its member miles points on the amount of time they spent playing with their cat or dancing at home in virtual live concerts in order to make it a fun experience. According to Thai Airways, a high number of new members registered through the mobile app and have been rewarded miles points. Through this campaign, Thai Airways presented themselves as a concerned airline that wanted their members to stay at their home during the pandemic, but also to have hope for better future travel by redeeming their miles points.

The spokesperson from Thai Airways has announced that they will be organising a webinar meeting at 11:00 am on 25th April 2020 to provide more information about the campaign. Prepare the spokesperson's introductory speech, as well as answers to the doubts and queries that could be raised during the webinar regarding the functioning and credibility of the marketing programme.

First part for the Influencer team

Thai Airways is the national air carrier of Thailand, a popular Southeast Asian country. The airline is well-known for its hospitality, catering service, cooperative staff and hygienic cabin. Royal orchid Plus is Thai Airways' membership programme that awards miles points to its members. The programme has more than two million registered members.

In 2020, the airline suffered heavy losses due to the global Covid-19 pandemic, which resulted in total lockdown of the country. The tourism industry suffered a massive downfall in 2020, as did the aviation industry. Amidst the losses the airline faced, the airline team kept working towards solutions to retain its image among its members as well as on new potential Influencers.

On 24th April 2020, the Airline, with the collaboration of Wunderman Thomson, launched a creative campaign – Stay Home Miles Exchange programme and posted on various social media platforms. The theme of the campaign is exact reversal of the traditional miles points programmes in which members gain miles points depending on the class they fly and frequency of their air travel. The campaign focussed on requesting that people stay at home and be rewarded miles points every four hours. This programme was exclusively for registered Thai Airways Royal Orchid Plus members. In addition, new members could be added by simply registering on their website. The members needed to install a mobile application, and automatically their movement within a 100-metre radius was tracked. The app also rewarded its members miles points on the amount of time they spent playing with their cat or dancing at home in virtual live concerts in order to make it a fun experience.

The campaign lasted until 23rd May and collected miles would then transfer to members' Royal Orchid Plus account by 10th July 2020. Through this campaign, Thai Airways presented themselves as a concerned airline that wanted their Influencers to stay at their home during the pandemic, but also to have hope for better future travel by redeeming their miles points.

The spokesperson from Thai Airways has announced that they will be organising a webinar meeting at 11:00 am on 25th April 2020 to provide more information about the campaign. As the representatives of global travellers, the Social Travel Influencers need to get prepared for the meeting.

Second part for the Airline team

On 23rd May 2020, the marketing campaign ended for android users, and on 26th May 2020 for iOS users. The airline witnessed huge interest among people regarding its campaign. On 10th July 2020, the Royal Orchid Plus accounts of registered members were credited with miles points to be redeemed once they start travelling.

The creative marketing campaign of the airline won several awards in different categories. The campaign won Lotus Grande in the Direct section and Gold in the Brand experience section at ADFEST 2021. It also won Grand Prix, Grand Prix and Gold in Mobile, Brand Experience & Activation, and Brand Experience & Activation + Direct + Mobile, respectively, in Spikes Asia 2021.

The airline benefotted from the positive response to the marketing campaign, which resulted in 37,230 registered users, 8,09,152 miles awarded and 32,36,608 million hours spent at home. The mobile app has come #1 in the health and fitness category in the app store with a US$764K PR value.

After seeing the popularity of the marketing campaign, Thai Airways came up with another marketing campaign – Destination Menu – where dishes from 12 destinations can be delivered to its members who crave travel and local foods from famous destinations.

The spokesperson will hold another webinar meeting on 11th July 2020 at 11.00 am. Prepare the spokesperson's speech regarding the success of the campaign and future plans of the airline, as well as answers to potential queries.

Further reading

De Alwis, A., & Andrlic, B. (2016). Social media in destination marketing. *International Journal of Management and Applied Sciences*, 2(January), 121–125.

Kruja, D. (2018). Destination marketing research. In D. Gursoy & C.C. Chi (Eds.). *The Routledge Handbook of Destination Marketing* (pp. 35–48). New York: Routledge.

Marome, W., & Shaw, R. (2021). COVID-19 response in Thailand and its implications on future preparedness. *International Journal of Environmental Research and Public Health*, 18(3), 1–11. https://doi.org/10.3390/ijerph18031089

Spikes Asia. (2020). Stay Home Miles Exchange. https://www2.spikes.asia/winners/2021/brand_experience/entry.cfm?entryid=2059&award=99&order=3&direction=2

Thai Airways. (2020). Thai Stay Home Miles Exchange. www.thaiairways.com/sites/en_TH/rop/Promotion/THAI-quarantine-miles.page

Wunderman Thompson. (2020). Thai Airways Stay Home Miles Exchange. www.wundermanthompson.com/work/thai-stay-home-miles-exchange

Case 10

COMMUNICATING HEALTHY FOREST AIR AS MEDICINE

Maurizio Droli and Yasuo Ohe

Duration

Case execution can last from a minimum of 120 minutes to a maximum of 360 minutes, depending on the degree of complexity of task assignments. The complexity level of task assignments can be 'low' (LC), 'medium' (MC) and 'high' (HC). The accomplishment of each task level requires 120 minutes of work, as will be explained in the 'Case' section.

Learning objectives

Upon completing the case, participants will be able to:

- explain the importance of forested areas for health-tourism innovation purposes
- communicate the key resources needed to organise forest therapy practices
- discuss 'forest bathing' and 'forest therapy' as opportunities to innovate wellness tourism and health tourism, respectively.

Based on the definition of Forest Therapy (Forest Therapy Society, 2021a), the establishment of a 'Forest Therapy Station' (FTS) requires the following steps: (1) the execution of controlled/randomised clinical trials; (2) the assessment of the health effects produced by experiencing forest paths; (3) the anonymous peer revision of the study; (4) its publication as a scientific contribution; and (5) the establishment of a medical-driven managerial unit to run forest therapy activities on a year-round basis. Thus, this case study does not deal with the establishment of a 'Forest Therapy Station' or of a 'Forest Therapy Path'.

After completing the case, students will be able to prepare an early version of the business model (Osterwalder et al., 2011) that is referred to as the 'Forest Bathing

DOI: 10.4324/9781003182856-10

Station' (FBS). For the purposes of this chapter, an FBS may be described as a place in which forest bathing services have been made available to people (residents, one day visitors and tourists), allowing them to appreciate the psychological benefits ('valuability'), rarity, inimitability and low-substitutability of forests through the experiencing of wellness, recreational, educational, cultural and other sustainable outdoor activities.

The lessons learned are in line with the 'World Health Organization Manifesto for a healthy and green recovery from COVID-19', stressing the importance of preserving natural resources for both human health and economic development purposes.

Target audience

This case focuses on value creation for customers that lies at the core of marketing. It therefore requires students to have a prior knowledge of tourism marketing. Moreover, the concepts of 'forest bathing' and 'forest therapy' represent ecosystem services. As such, discussing the case requires students to have a basic knowledge of forests, ecosystem services and forest ecosystem services for human health and wellness purposes. Last but not least, the case applies the so-called 'Resource Based Theory' (RBT) (Barney et al., 2011; Penrose, 2009; Wernerfelt, 1984), focussing on the important role played in economic competition by resources capable of generating 'above average results' (thus 'valuable'), and that are concentrated among competitors ('rare'), lowly imitable and lowly substitutable resources. As such, students dealing with the case should have a prior knowledge of RBT. The case is applicable to students at different levels of education (bachelor's, master's, PhD, and vocational training).

Teaching instructions

The case is designed for group work. There need to be five groups working separately, each with a minimum of three and a maximum of five students. Each of those five groups of students should be assigned to one of the following five research-fields.

Field 1: Valuability of forest-based experiences for human health purposes

The first group of students should review the scientific literature to identify the differences between experiencing generic forests and experiencing Relatively Clean and Healthy Air (RCHA) at forest therapy bases. Participants can classify these benefits according to the type of natural resource determining them, such as mountain forest, urban forest and coastal forest. Other students can then break down these benefits according to the type of people who have obtained the benefit, for example based on age or health condition (pathological/healthy patient).

Field 2: Product 'rarity'

The second group of students could analyse the spatial concentration (rarity) of places, especially forests, in which those benefits were obtained. These observations should allow students to identify the locations where the benefits gave rise to original and consistent tourism products. In the specific case of 'forest bathing', students should observe what is proposed by countries such as Japan, which leads the world in studies on the health effects obtained by immersion in forests, as well as South Korea and neighbouring European countries. Students should then recognise how conducting a clinical study that illuminates the health benefits of using a specific forest trail is not required for developing a new Forest Bathing Station. Furthermore, students could discuss the similarities and the differences that lie in the marketing of forest bathing initiatives compared to the sustainable marketing of those initiatives.

Field 3: Product 'inimitability'

The third group of students should identify the most 'socially complex' (thus less imitable) organisations running forest-based care initiatives. The existence of a tourism product that is difficult to imitate çan help the organisations promoting it to achieve a defensible competitive advantage, following historically rooted managerial theories. Students should collect information on the networks marketing forest bathing initiatives to both the final and intermediate (B2B) health/wellness-tourism market.

Field 4: Product 'low-substitutability'

Tourism products which are lowly imitable, *and* difficult to be substituted with other equivalent ones offer more possibilities to the organisations marketing them to achieve a defensible competitive advantage.

One of the less-substitutable, therefore most strategically relevant, natural resources existing at the preselected forest is the quality of the air, especially for forest bathing and forest therapy purposes. In other words, rural areas characterised by huge forest heritage may achieve different levels of air quality depending on their proximity to urban areas, their geographical position with respect to the dominant winds, altitude, etc. This fourth group of students should assess the air quality found at each preselected forest.

Field 5: Marketing organisation

The fifth group of students should observe the marketing strategies, techniques and tools that have been adopted in places sustainably leveraging forests as strategic resources for the purposes of developing them as health tourism destinations. This group should consider forest-based care initiatives to observe the key resources, key segments, key activities and other key elements of the business model adopted by those organisations. In this way, students could be helped to recognise the importance of RCHA for both wellness tourism and health tourism development purposes.

As a second step, students working on any research field should generate hypotheses of marketing initiatives leveraging forest aerosols. The third step is the development of a draft scheme of the business model for a tourism destination to leverage forest aerosols and/or other natural resources as 'health devices' for both wellness tourism and health-tourism marketing purposes.

Teaching methods and equipment

Alternative teaching methods can be employed to develop this case. Desk analysis should be performed, for which individual work is required. Having gathered the required information, group work, role playing and game plays should allow students to become aware of the potentials of healthy air for sustainable marketing purposes. Following this, both virtual and on-site company and/or destination visits could be organised. During those visits, the method of 'participant observer' can be enacted, and further information, both qualitative and quantitative, collected. Students could then be encouraged to adopt 'story telling' to describe the impact of experiencing forests on tourists' satisfaction levels. In-forest experiments, and audio and video-recordings could also be created.

This case study focuses on the application of scientific evidence generated by medical studies for sustainable marketing purposes. It stimulates the students' ability to work with people possessing different skills. The equipment required for this case comprises mobile phones, video cameras, computers, projectors, and flipcharts or white boards to synthesise and share the results obtained.

Case

Field 1: 'Valuability' of forest-based experiences for human health purposes

Forests have multiple functions. According to the Food and Agriculture Organization of the United Nations (FAO), forests can be defined as a land that 'overcomes the threshold of 10% minimum crown cover, including both natural forests and forest plantations and excluding stands of trees established primarily for agricultural production, e.g. fruit tree plantations' (FAO, 2000).

Benefits produced by nature on human health originate from the innate human attraction to other living organisms, which is in accordance with Wilson's 'biophilia' hypothesis (1984). The well-being and healing effects of forest-related activities have been variously elucidated by scholars, particularly in Northern European countries (e.g. Hjalager & Flagestad, 2012). Nevertheless, forest therapy practices based on volatile organic compounds (VOCs) differ from experiencing forests, nature and the outdoors as a whole (Kneipp, 2010).

Consistently, the physiological and psychological beneficial effects of forest-based experiences have been investigated by adopting different medical and psychological lenses.

The main natural resources whose potentials are harnessed for health-tourism purposes are forest aerosols, more specifically VOCs, produced by both conifers and hardwoods to defend their plant mechanisms against the environment and herbivores. 'Terpenes' represent the largest class of organic VOCs produced by various plants and are one of the main constituents of forest aerosols. They include α-pinene, camphene, limonene, camphor, cymene and other forms of isoprenoid compounds that have traditionally been used to treat various illnesses, despite a lack of specific knowledge regarding the therapeutic mechanisms triggered by them. In fact, terpenes, mainly produced by conifer trees, have only recently been classified according to their anti-inflammatory, neuro-protective, or anti-tumorigenic properties (Cho et al., 2017).

With the aim of optimising the healing benefits deriving from both inhaling VOCs and from experiencing forests, the original concept of 'forest bathing' (in Japanese: 'Shinrin-yoku', 森林浴) was coined. The concept can be described as: 'visiting forests and woods for relaxation and recreation purposes, while breathing in volatile organic substances, called phytoncides (wood essential oils), which are antimicrobial VOCs derived from trees' (Li, 2010, p. 9).

Revisions of studies highlighting the positive effects deriving from experiencing forest immersions include: Hansen et al. (2017) Kamioka et al. (2012), Lee et al. (2017), Li (2019), Rajoo et al. (2020) and Twohig-Bennett and Jones (2018).

Based on this evidence, the concept of 'forest therapy' has been suggested to indicate 'a research-based practice supporting the healing of individuals through their immersion in forest atmosphere' (Forest Therapy Society, 2021a).

This working group should focus on the health benefits obtained from experiencing forests. The tasks assigned could encompass the following:

- *Low Complexity (LC)*: identifying and citing the primary studies that highlight the main health benefits obtainable from experiencing forests.
- *Medium Complexity (MC)*: breaking down these benefits according to the type of forest (urban forests, mountain and/or coastal forests).
- *High Complexity (HC)*: preparing a brief report on the results obtained.

To accomplish this task assignment, Google Scholar or other repositories of scientific studies could be used.

Field 2: Product 'rarity'

Japan's forests cover some 25 million hectares (Ministry of Agriculture, Forestry and Fisheries of Japan, 2021a, 2021b), most of which were planted after World War II. Of this planted forest, 51% has reached the optimal point for utilisation according to different multifunctional uses, the value of which, in monetary terms, has been estimated at some 70 trillion yen. The healing potentials of those forests started to be harnessed through the development of 'forest therapy bases' (FTBs) in selected forested areas in 2006. FTBs have been described as 'scientifically approved locations with forests and urban environments where experiments were conducted analysing the healing and preventive medical benefits due to the environment' (Forest Therapy Society, 2021b). Recently, FTBs established in Japan have been compared with those established in other countries, e.g. Italy (Droli et al., 2020).

This working group should focus on issues of forest quality and accessibility. The tasks assigned for this working group should encompass the following:

- *LC*: shortlisting a minimum of three and a maximum of ten forested rural-tourism destinations, located at any altitude, accessible by train and potentially suitable for hosting an FBS.
- *MC*: analysing travel costs from each forest to the university campus / municipality of students' residence / most populated city within a two-hour train journey.
- *HC*: locating the one to three forests accessible at the lowest travel costs/ shortest travel time.

Field 3: Product 'inimitability'

Local history and productive traditions can help tourism destinations to develop lowly imitable tourism products through the generation of historically rooted capabilities. Japan has been saving its forests from the rapacious deforestation that occurred over the last 300 years through various programmes, including the development of silviculture, forest restoration and community forest management (Marten, 2005).

Some of the most important steps through which Forest therapy has been established are the following:

1. *Technical insight by a technologist (1982):* Notwithstanding it was a term already in existence, the use of 'shinrin-yoku' ('forest bathing trips') to refer specifically to a form of forest therapy was introduced by Mr. Akiyama Tomohide, Director General of the Japanese Forestry Agency (JFA), and by the Ministry of Agriculture, Forestry and Fisheries in the national context in 'Showa 57' (Li, 2010; Takayama, 1982).
2. *Initiation gap (1983–2003):* Following the launch of the concept of 'forest bathing trips' in 1982, several studies on the positive effects of forest bathing activities were published at a local level. However, the data gathered by those studies was heterogeneous and not comparable, and thus not useful for benchmarking purposes. Furthermore, there was no consensus on what standard evaluation criteria and proxies should be adopted by such studies in order to identify forests suitable for becoming 'Forest Therapy Stations'. This is important, as 'forest bathing', which has evolved into 'forest therapy', necessarily requires the healing effects of forests to be quantifiably demonstrated through scientific publications (Lee et al., 2012).

3. *Opportunity recognition (2004–2006):* In 2003, the Japanese Forest Agency and the Forest Products Research Institute (FPRI) recognised the need to assess specific forest bathing development targets and began discussing a new 'Plan of the Forest Therapy Base'. Consequently, a number of different research projects were undertaken. In 2004, the Ministry of Agriculture, Forestry, and Fisheries of Japan started to support a specific research project entitled 'Physiological Effects of Forest Environmental Components on Humans' to investigate the therapeutic effects of forests on human health (Tsunetsugu et al., 2010, p. 32). In 2004, a forest therapy research group was instituted to provide an 'illustration of the physiological effects of the forest element on human subjects'. This research group involved Professor Qing Li of the Nippon Medical School in Tokyo as Principal Investigator, the Forest Research Institute (FRI), and Chiba University through funding made available by the Japanese Ministry of Agriculture, Forests and Fisheries (International Society of Nature and Forest Medicine, 2021a). The team gathered sufficient evidence concerning the relation between forest therapy practice and human health improvement.
4. *Formal evaluation (18th April 2006):* The Forestry Agency, the National Land Afforestation Promotion Organization (FANAPO) of Germany, and the Forestry Research Institute certified the first six 'forest therapy bases' (Ohe et al., 2017).
5. *Formation of a project improvement team (2007–2011):* In 2007, the academic scholar-based Japanese Society of Forest Medicine (SFM) was created to promote scientific research into the benefits of forest medicine, including the effects of forest bathing trips and the therapeutic effects of forests on human health, to national universities, academic institutes, local governments and private companies (SFM, 2008). Professor Qing Li of the Nippon Medical School is the President of the Association, and served as a principal investigator in its early academic research on the issue. The Society has implemented a great number of internationally recognised studies on forest medicine over the last 20 years. In 2008, the National Conference on the Forest Therapy Base Network, the Forestry Agency, the Forestry Research Institute, Forest Medical Research (FMR), and the Forest Therapy Society (FTS) non-profit organisation stipulated a new framework for promoting forest therapy and accelerating various projects related to the development of forest therapy across different prefectures. In 2011, the International Society for Nature and Forest Medicine (INFOM) was founded, with the first Symposium of INFOM, co-organised with the JSFM, being held at Kyoto University in 2012. The chairpersons were Qing Li (Nippon Medical School), Yoshifumi Miyazaki (Chiba University) and Takahide Kagawa (Forestry and Forest Products Research Institute of Japan). INFOM's main objectives include promoting natural and forest medicine, spreading awareness of its advantages across western, eastern and traditional (conventional) medicine, and facilitating the development of even more effective and sustainable medical therapies (INFOM, 2017). These various concepts are protected under trademark by the FTS.

This working group should focus on potential competitors and/or strategic partners. The tasks assigned to this working group could encompass the following:

- *LC*: shortlisting a minimum of three and a maximum of ten 'forest bathing' experiences operating within the municipality, region, country of interest, or at an international level.
- *MC*: closely describing those experiences in regards to Products sold to the market, Prices, Places and the other P's of marketing.
- *HC*: reporting on the results obtained.

Field 4: Product 'low-substitutability'

Within the field of tourism research, the offer of forest therapy services has been discussed as a niche market, and one not suitable for mass tourism (Hamdan & Low, 2014). Similarly, the need to avoid forest overcrowding so as to preserve forest ecosystems has been described (Gallagher & Hermann, 2014), as has the possibility of leveraging the authenticity of Japanese forests to attract outbound tourism (Carlisle et al., 2013). Another recent study defined 'forest therapy tourism' as 'an emerging type of rural tourism trying to internalise forests' multifunctionality, focusing on relaxation effects and income creation opportunities' (Ohe et al., 2017, p. 323). This current study derives the concept of forest therapy tourism from the prevailing health and wellness aims of both 'forest bathing' and 'forest therapy' activities. Following these aims, forest therapy tourism can be represented as a forest atmosphere based health tourism product that is capable of harnessing unexpressed potentials of healthier rural areas.

As such, the working group should focus on the quality of air and its assessment. The tasks assigned for this working group could encompass the following:

- *LC*: consulting internet platforms providing data on air pollution that are managed by public authorities to identify the relevant pollutant(s) to monitor (at least Particulate Matter PM 2.5 and PM 10) through an in-field campaign. Assessing the levels of air pollution during a day of low/no-wind in the most populated city within range of a two-hour car journey from the FBS / the students' city of residence / the university campus, etc.
- *MC*: assessing the air pollution levels (same data) in the forests suitable to host the FBS.
- *HC*: compiling a report illustrating the air quality, and whether a 'healthy air' (evidence-based) competitive advantage exists at the FBS and, if so, its entity.

For this, a low-cost system of air pollution monitoring should be adopted (Rogulski, 2017).

Field 5: Marketing organisation

A specific blueprint aimed at facilitating the establishment of a Forest therapy (Shinrin) Station has been defined (Miyazaki, 2017).

Starting from the first six FTBs to be certified in 2006, as of January 1st 2021, Japan now boasts one 'highly recommended (2 stars) base' at Shinano-Machi (Nagano), and some 63 certified FTBs, including in Hokkaido (2), Tohoku (6), Kanto (14), Hokuriku–Koshinetsu (17), Tokai (3), Kansai (4), Chugoku (6), Shikoku (2), Kyusyu (6) and Okinawa (1), according to the Forest Therapy Society (FTS, 2021c). The achievement of 100 certified FTBs has been foreseen by 2050.

FTBs are jointly marketing the following key-elements:

1. *The forest therapy base certification*: The setting-up of an FTB requires self-responsible players utilising comparable scientific results. As previously suggested, 'forest bathing' activities must demonstrate their healing effects in order to be recognised as forest therapy activities (Miyazaki et al., 2014). For this purpose, the Japanese Society of Forest Medicine (SFM) has initiated a patented approach to assessing the therapeutic effectiveness of forest bathing and a method for the measurement of corresponding physiological benefits, while the National Forestry Agency (NFA) has instituted the first 'Forest Therapy Physiological Experiment Example' worksheet to augment the comparability of clinical experiment results. To increase the self-responsibility of decision markers, the FTS has established specific quality standards for forest therapy, forest therapy roads and experimental contents (FTS, 2021a, 2021b). In addition to positive ('above threshold') results obtained from physiological and psychological experiments, other 'non-medical' evaluation criteria used for FTB certification purposes include: (i) natural social conditions (good environmental conditions, environmental improvements and facilities, access); and (ii) accommodation facility aspects, such as a management unit, dedicated menu, acceptance of local residences, future plans and a sales point.
2. *Forest therapy roads*: These represent the physical place within an FTB where the therapeutic effects of forest bathing activities have been scientifically proven. Superficially, forest therapy roads consist of a path, generally wider than a normal urban sidewalk, that is far removed from artificial noise, odours and colours. Very often, they are planned to follow gentle slopes while offering completely barrier-free environments and fully guaranteeing accessibility to wheelchair users. Nevertheless, some areas in an FTB can be populated by trees whose VOCs, on inhalation, have limited or no proven clinical effects (Surette et al., 2013). 'Therapy roads' are required to assess those obstacles for sustainable health and wellness purposes.
3. *Forest therapy guides*: Guests welcomed by an FTB may either be healthy or be suffering from a medical condition. Forest therapy guides must therefore possess a range of interdisciplinary competences. Their main task is to provide guests with advice on how to improve the effect of the forest bathing experience on their health and wellness. Additionally, they must possess basic scientific knowledge of both the environment and of human physiology in order to be able to provide general information on the value of natural resources and illustrate the healing effects of the forest, respectively. Most importantly, they

are required to select safe and secure walking paths and roads appropriate to the individual guest by evaluating the general conditions, the visitors' age and health conditions and the specific health-improvement targets to be achieved through forest therapy activities. A 'forest therapy guide' is also required to help guests experience the enveloping comfort of the forest, facilitating the harmonisation of human beings with the rhythms of the natural environment. Together, forest therapy guides and forest therapists represent two of the most important professional groups within the broader forest therapy station/base (Miyazaki, 2017).

4. *Forest therapists*: 'Forest therapists' must necessarily be medical doctors, while forest therapy guides may also be medical doctors, according to the guidelines established by the NPO Forest Therapy Society. The main task of forest therapists is to plan and provide appropriate programmes for their guests. Furthermore, they provide forest therapy guides with training on effective therapeutic activities. In order to do so, they must have the knowledge of a forest therapy guide, specialised knowledge on health and psychology and possess high-level communication skills.
5. *Forest therapy community events*: Special events may contribute to an even stronger and more competitive public-private partnership in this field. In 2015, after registering 'Walking Forest Therapy Roads' as a trademark, the International Society of Nature and Forest Medicine (INFOM) started organising a special event called 'Walking Forest Therapy Roads with a Doctor' with the help of a subsidy from the Green and Water Forest Fund created under the auspices of the National Land Afforestation Promotion Organization (INFOM, 2017).

This working group should comprise all the students participating in the case study. Task assignments should encompass the following:

- *LC*: bringing together the work done by each research group through the preparation of a video presentation synthesising the information made available by all the various previously compiled group reports.
- MC: presenting the report to the members of the other research groups, sharing the results, and formulating conclusions.
- *HC*: correctly setting up the business model (Osterwalder et al, 2011) for each FBS.

Conclusion

In conclusion, the promotion of scientific studies demonstrating the benefits produced by breathing 'relatively clean and healthy air' should increase the market visibility of a tourism destination. Students should track, manage and capitalise that added visibility whilst also involving local communities and protecting the environment.

Acknowledgements

The authors wish to thank Dr. Gianluca Benatti at C2Partners, Milan, for kindly suggesting a way in which to make the title of this chapter a little less technical and a little more engaging.

References

Barney, J. B., Ketchen Jr., D. J., & Wright, M. (2011). The future of resource-based theory: Revitalization or decline? *Journal of Management*, 37(5), 1299–1315. doi:10.1177/0149206310391805

Carlisle, S., Kunc, M., Jones, E., & Tiffin, S. (2013). Supporting innovation for tourism development through multi-stakeholder approaches: Experiences from Africa. *Tourism Management*, 35, 59–69. doi:10.1016/j.tourman.2012.05.010

Cho, K. S., Lim, Y., Lee, K., Lee, J., Lee, J. H., & Lee, I. S. (2017). Terpenes from forests and human health. *Toxicological Research*, 33(2), 97–106. doi:10.5487/TR.2017.33.2.097

Droli, M., Gervasio-Radivo, G. G., & Iseppi, L. (2020). Does the establishment of a 'Forest Therapy Station' in a low-mountain mixed hardwood forest make sense? In C. Bevilacqua, F. Calabrò, & L. D. Spina (Eds.), *International Symposium: New Metropolitan Perspectives* (pp. 67–79). Switzerland: Springer, Cham.

Food and Agriculture Organization of the United Nations. (2000). Global Forest Resource Assessment, 2000. Terms and Definitions. www.fao.org/3/Y1997E/y1997e08.htm#TopOfPage.

Forest Therapy Society. (2021a). Nature contributes to mental and physical health. www.fo-society.jp/therapy/cn45/index_en.html.

Forest Therapy Society. (2021b). 62 forests across Japan. www.fo-society.jp/quarter/cn49/62forest_across_japan.html.

Forest Therapy Society. (2021c). Certified forests. https://fo-society.jp/en/forests.html.

Gallagher, D., & Hermann, N. (2014). Antiepileptic drugs for the treatment of agitation and aggression in dementia: Do they have a place in therapy? *Drugs*, 74(5), 1747–1755. doi:10.1007/s40265 014-0293-6

Hamdan, M., & Low, K. C. P. (2014). Ecotourism development in Brunei Darussalam. *Transnational Corporations Review*, 6(3), 248–272. doi:10.5148/tncr.2014.6304

Hansen, M. M., Jones, R., & Tocchini, K. (2017). Shinrin-Yoku (forest bathing) and nature therapy: A state-of-the-art review. *International Journal of Environmental Research and Public Health*, 14(8), 851. doi:10.3390/ijerph14080851

Hjalager, A.-M., & Flagestad, A. (2012) Innovations in well-being tourism in the Nordic countries. *Current Issues in Tourism*, 15(8), 725–740. doi:10.1080/13683500.2011.629720

International Society of Nature and Forest Medicine. (2021a). Introduction. www.infom.org/aboutus/introduction.html.

International Society of Nature and Forest Medicine. (2017). Completion report 'Walking Forest Therapy Roads® with a Doctor' in 2014 operation year. http://infom.org/news/2015/03/walkingforesttherapyroads.html.

Kamioka, H., Tsutani, K., Mutoh, Y., Honda, T., Shiozawa, N., Okada, S., … & Handa, S. (2012). A systematic review of randomized controlled trials on curative and health enhancement effects of forest therapy. *Psychology Research and Behavior Management*, 5, 85–95. doi:10.2147/PRBM.S32402

Kneipp, S. (2010). *Meine Wasserkur: So sollt ihr Leben*. Stuttgart: Verlag.

Lee, I., Choi, H., Bang, K. S., Kim, S., Song, M., & Lee, B. (2017). Effects of forest therapy on depressive symptoms among adults: A systematic review. *International Journal of Environmental Research and Public Health*, 14(3), 321. doi:10.3390/ijerph14030321

Lee, J., Park, B. J., Tsunetsugu, I., & Miyazaki, Y. (2012). Forests and human health – Recent trends in Japan. In Q. Li (Ed.), *Forest Medicine* (pp. 243–257). New York: Nova Science Publishers.

Li, Q. (2010). Effect of forest bathing trips on human immune function. *Environmental Health and Preventive Medicine*, 15(1), 9–17. doi:10.1007/s12199-008-0068-3

Li, Q. (2019). Effets des forêts et des bains de forêt (shinrin-yoku) sur la santé humaine: une revue de la littérature [Effect of forest bathing (shinrin-yoku) on human health: A review of the literature]. *Sante Publique*, 13, 135–143. doi:10.3917/spub.190.0135

Marten, G. (2005). Environmental tipping points: A new paradigm for restoring ecological security. *Journal of Policy Studies*, 20, 75–87.

Ministry of Agriculture, Forestry and Fisheries of Japan. (2021a). Multifunctional roles of forests. www.rinya.maff.go.jp/new/hakusyoeigo/english18/textp2.htm.

Ministry of Agriculture, Forestry and Fisheries, Japan. (2012b). Forestry Agency's Annual Report on Forest and Forestry in Japan, Fiscal year to March 31, 2012. www.rinya.maff.go.jp/j/kikaku/hakusyo/24hakusyo/pdf/h24summary.pdf.

Miyazaki, Y. (2017). Science of Natural Therapy. www.marlboroughforestry.org.nz/mfia/docs/naturaltherapy.pdf.

Miyazaki, Y., Ikei, H., & Song, C. (2014). Forest medicine research in Japan. *Japanese Journal of Hygene*, 59(2), 122–135. doi:10.1265/jjh.69.122

Ohe, Y., Ikei, H., Song, C., & Miyazaki, Y. (2017). Evaluating the relaxation effects of emerging forest-therapy tourism: A multidisciplinary approach. *Tourism Management*, 62, 322–334. doi:10.1016/j.tourman.2017.04.010

Osterwalder, A., Pigneur, Y., Oliveira, M. A. Y., & Ferreira, J. J. P. (2011). Business Model Generation: A handbook for visionaries, game changers and challengers. *African Journal of Business Management*, 5(7), 22–30.

Penrose, E. (2009). *The Theory of the Growth of the Firm*. New York, US: Oxford University Press.

Rajoo, K. S., Karam, D. S., & Abdullah, M. Z. (2020). The physiological and psychosocial effects of forest therapy: A systematic review. *Urban Forestry & Urban Greening*, 54, Article 126744. doi:10.1016/j.ufug.2020.126744

Rogulski, M. (2017). Low-cost PM monitors as an opportunity to increase the spatiotemporal resolution of measurements of air quality. *Energy Procedia*, 128, 437–444. doi:10.1016/j.egypro.2017.09.026

Society of Forest Medicine. (2008). Introduction to the Japanese Society of Forest Medicine. http://forest-medicine.com/epage01.html.

Surette, S., Vanderjagt, L., & Vohra, S. (2013). Surveys of complementary and alternative medicine usage: A scoping study of the paediatric literature. *Complementary Therapies in Medicine*, 21, S48–S53. doi:10.1016/j.ctim.2011.08.006

Takayama, Y. (1982). Blueprint for the shinrin-yoku. The Asahi Shimbun, July 29.

Tsunetsugu, Y., Park, B. J., & Miyazaki, Y. (2010). Trends in research related to 'Shinrin-yoku' (taking in the forest atmosphere or forest bathing) in Japan. *Environment Health Preventive Medicine*, 15(1), 27–37. doi:10.1007/s12199-009-0091-z

Twohig-Bennett, C., & Jones, A. (2018). The health benefits of the great outdoors: A systematic review and meta-analysis of greenspace exposure and health outcomes. *Environmental Research*, 166, 628–637. doi:10.1016/j.envres.2018.06.030

Wernerfelt, B. (1984). A resource-based view of the firm. *Strategic Management Journal*, 5(2), 171–180.

Wilson, E. O. (1984). *Biophilia*. Cambridge, Massachusetts, US: Harvard University Press.

Further reading

Choukas-Bradley, M. (2020). *Resilience: Connecting with Nature in a Time of Crisis*. John Hunt Publishing.

Li, Q. (2018). *Shinrin-Yoku: The Art and Science of Forest Bathing*. London, UK: Penguin Life.

Shin, W. S., & Lee, J. (2020). Forest therapy. In N. J. Harper & W. W. Dobud (Eds.), *Outdoor Therapies* (pp. 159–172). New York: Routledge.

Harper, N. J., Rose, K., & Segal, D. (2019). *Nature-Based Therapy: A Practitioner's Guide to Working Outdoors with Children, Youth, and Families*. Canada: New Society Publishers.

Case 11

VIRTUAL AND AUGMENTED REALITY IN TOURISM

Dare to research?

Patricia Isabel Santateresa-Bernat

Duration

The case is intended for 12 sessions, around 90 minutes each.

Learning objectives

Upon completing the case, participants will be able to:

- identify the reliable secondary sources of information about the use of virtual and augmented reality in a tourist destination
- determine the specific objectives of a research project about habits, attitudes and behaviour concerning the use of the virtual and augmented reality at the tourism destination
- write a report establishing the starting research hypotheses; this must follow the APA style guidelines
- design a structured survey fulfilling the theoretical structure and using the most adequate questions and scales to fulfil the research objectives through digital tools
- learn how to perform a pollster's tasks through fieldwork
- develop a database with questionnaire responses
- apply univariable and bivariable analyses with the PSPP program
- prepare an analysis report to discuss the major findings, conclusions and recommendations to destination management organisations to make decisions about the implementation of virtual and augmented reality in their destination.

DOI: 10.4324/9781003182856-11

Target audience

The case study is defined for those students that need to develop research skills in the field of tourism, representing bachelor's, master's, Ph.D. or vocational training who have a subject or a project that requires them to know how to investigate a topic. For the students, it is necessary to have knowledge about tourism marketing and statistical concepts to follow this case study. On the one hand, the procedures for the tourism market research require students to know about marketing to understand the importance of this kind of research to support the DMO decision taking. On the other hand, it would be useful if students had some knowledge of statistics at an introductory level, including descriptive statistics and basic tests like Chi-Square and the analysis of variance (ANOVA). It would also be helpful to have some knowledge of the field of tourism digitalisation, like virtual reality, augmented reality and how to use digital collaborative tools such as Google Drive.

Teaching methods and equipment

The case study consists of research work divided into a series of activities, with the purpose of developing the tourist market research to solve the problem to be studied, i.e. to know about the habits, attitudes and behaviour concerning the use of virtual reality and augmented reality in tourist destinations. These activities are scheduled in five phases that are part of the final project on the subject, so the students confront each of the activities as objectives to achieve in the short term, but the deadline for the project is at the end of the semester. The students work in teams of three to five members, and then consult secondary information sources about the virtual and augmented reality in journals, databases and other physical or web sources. The students are required to use digital collaborative tools to work together on each part of the report. In this way, the market research report is made up with the aim of responding to a problem with the subject instructors guiding and coordinating the work. Concerning the quantitative approach, in order to design the survey it is essential to access Google Forms; and software such as PSPP is necessary to make the statistical analysis.

The proposed for each phase is as follows:

Phase I

- First session: the task is proposed, the teams are configured and students start to look for further information to define the objectives and hypothesis.
- Second session: the students share their progress, get feedback and continue with their task to prepare the partial report regarding background, objectives and hypothesis.

Phase II

- First session: the theoretical framework about quantitative research is explained, the task is proposed and the students develop a Google form survey to fulfil the objectives.
- Second session: the students share their progress, get feedback and continue with their task.

Phase III

- First session: the students are informed on how to do the fieldwork by giving instructions as if they were pollsters.
- Second session: outside the class, the students have time to develop fieldwork and find their sample.

Phase IV

- First session: data preparation process with the students and explanation on how to use PSPP for univariable and bivariable analysis as frequencies distribution, descriptive measures, cross-tabulations, Chi-square and ANOVA.
- Second session: the students share their progress, get feedback and continue with their task to analyse the database to fulfil their research objectives.
- Third session: interpretation of the statistical measures to verify the hypothesis; and the students start to elaborate their partial analysis report.
- Fourth session: the students share their progress, get feedback and continue with their task.

Phase V

- First session: the final report and presentation is prepared with the findings, conclusions and recommendations to DMOs to use virtual and/or augmented reality in their destination.
- Second session: presentation of the outcomes of each team.

For the students, it can be easier to confront each phase if the first part of the session is dedicated to giving a theoretical basis of the concepts related to the tasks. Also, it is essential to give the students time to develop this practical work while providing them with the necessary support. It can be beneficial for the teams to share their progress at the beginning of each session in order to correct deviations and give them feedback.

Teaching instructions

Below is the task description to be followed in each phase.

Phase I: Background and objectives of the research

Collect information from the secondary sources to explore the problem and define its fundamental background, i.e. to know the habits, attitudes and behaviour concerning the use of virtual reality and augmented reality in tourist destinations. The students should work in teams of three to five members to synthesise the background, and then proceed to define both the specific objectives of their research and their starting hypotheses related to the background. It is important in this phase to give support and feedback to the students to ensure they focus on the research correctly.

Phase II: Research methodology

Define a theoretical framework about quantitative approach to train students to design the methodology to obtain the information proposed in the specific objectives, which involves the design of a structured survey. The students develop a survey in Google Forms and test it. For example, the students design a survey that fulfils each specific objective using different kinds of structured questions as multiple-choice, dichotomous questions and scales, trying to follow the general ordering of questions in a survey: (1) qualifying questions, (2) introductory questions, (3) main questions, (4) psychographics, (5) demographics and (6) identification information.

Phase III: Fieldwork

Present the main concepts to define the sampling method; then, each student must do the fieldwork to obtain responses to the survey across the sample profile, through different contact methods, either personal or via the internet. A sample of at least 50 valid surveys is requested for each team. The students can use a nonprobability sampling technique, such as convenience sampling, judgmental sampling, quote sampling, or snowball sampling. However, other means can include probability sampling techniques, such as simple random sampling, systematic sampling, stratified sampling, or cluster sampling.

Phase IV: Data preparation and analysis

With the instructor's guidance, each team will prepare a database with all the survey responses, and then analyse them with the support of a PSPP statistical program. It is essential that the students know how to identify which kind of analysis is most appropriate to apply to answer each specific objective, and then interpret the statistical measures to verify the hypothesis. Analysis can be in the form of frequencies distribution, descriptive measures, cross-tabulations, Chi-square and ANOVA. For example, while a descriptive frequency analysis would be applied to find out respondents' habits, in order to obtain indicators such as percentage, valid percentage and cumulative percentage, to identify the respondents' attitudes, a descriptive analysis would be applied, involving indicators such as measures of location (mean, mode and median) and/or measures of variability (range, variance and standard deviation). Furthermore, to test the hypotheses, analyses such as cross-tabulation could be applied, which indicates the relationship between the variables analysed, obtaining statistics such as the Chi-square. Finally, an ANOVA analysis could be applied in order to understand the relationship between groups and its significance.

Phase V: Preparation of the report

The students will prepare a report consisting of the outcomes from each research phase, including conclusions and future lines of research, in relation to the results obtained. The main objective of the research must be examined against the background. Then, the major findings should be discussed, as well as conclusions and recommendations to DMOs, so as to help them to make decisions about the implementation of virtual and/or augmented reality in their destination. The report must contain at least a problem definition, an explanation about the approach to the problem, a description of the research design, a description of the data analysis, the results, a discussion about limitations and caveats and, finally, conclusions and recommendations.

Context

Each phase could become a unique case to develop in the class. The students just need to have an instruction to contextualise their task. The context can be provided by the instructor, for example, Phase II can become a single case if the instructor provides the students with some specific objectives and hypotheses to design the survey.

Case

Introduction

Emerging technologies, such as virtual reality and augmented reality, have influenced both tourism supply and tourists alike. The understanding of information technologies and devices, in the new age of the constantly evolving IT industry, is crucial (Beck, Rainoldi & Egger, 2019). Especially in tourism, it is necessary to follow and apply new information technologies to maintain the growth and popularity of the tourism market. Early information technologies influence the tourist market, companies, service providers, and, last but not least, travellers. Also, this indicates a corresponding transition in tourist behaviour when planning, arranging and experiencing travel. Due to this information, the importance of getting familiar with the latest high-tech IT inventions, which are usable in the tourism industry, is essential. The latest immersive IT technologies that appealed to the tourism industry were virtual reality and augmented reality (Yung & Khoo-Lattimore, 2019).

The introduction of virtual reality in tourism brought about a big impression on service suppliers and travellers during the period of planning and on-site experience. With the rise of virtual reality, it is anticipated that there will be increasing opportunities to use it in new aspects of travel and services that can increase the destination value from the tourist point of view (McFee et al., 2019).

Previous studies have also confirmed the positive feedback from travellers using virtual reality for planning or on-site exploration, education and experience. Furthermore, according to the results of previous research, virtual reality increases interest in visiting and re-visiting given sites. Owing to the development and popularity of virtual reality devices, especially among younger generations, we can assume that in the future, virtual reality devices will become the mainstream adaptation for the tourism industry. As a result of this information, the understanding and acquaintance with virtual reality technologies has proven vital for tourism. Additionally, it is important to understand the application of virtual reality to web travel portals, museums, historic sites, etc., so as to improve the tourist experience. Moreover, the service provider using the virtual reality devices must be acquainted with them in order to make their usage easily available and understandable for the tourists. Understanding and usage of virtual reality and/or augmented reality by destination management organisations will empower and deliver more advantages to the destination. Furthermore, it will attract substantial numbers of new tourists and enable people who have been unable to travel previously due to language barriers or disabilities to experience travel (Beck & Egger, 2019).

Therefore, it is necessary to carry out research that supports destination management organisations in understanding the level of use of virtual and augmented reality, and the predisposition of tourists to use virtual and augmented reality when visiting a destination.

Phase I: Background and objectives of the research

The main objective of the research is to know habits, attitudes and behaviour concerning the use of virtual reality and augmented reality in tourist destinations. Therefore, the case will consist of defining both the specific objectives of the topic and the starting hypotheses. For this, the students will look for information in external secondary sources (books, journals articles, rigorous internet portals, etc.) on the topic under investigation. This information will allow students to develop the background and define the specific objectives and hypotheses, and will also introduce the research presentation. The delivery of this report must meet the following guidelines:

- Overall, a group of three to five students should write an introduction regarding the topic, explaining the importance of knowing the use or value that the tourist gives to the use of virtual and augmented reality in destinations throughout their visit. On the other hand, it is important to explain how this information is useful to destination management organisations, in supporting them to make decisions about the implementation of virtual and/or augmented reality in their destination, because nowadays it is a tool that could increase the value of a destination to tourists. Also, it is needed to provide examples of applications in tourism destinations of virtual and augmented reality.

- In the second part of the report, the students should explain in a reasoned way the information gathered about the different habits, attitudes and consumer profile towards the use of this technology in a tourist destination. This information must be composed of truthful data published in various rigorous sources. Therefore, in the reasoned statement of the background, they should cite, in APA style, the sources consulted in the text.
- Once the background is set, the students should define specific objectives of the research, based on the information presented. They are required to propose at least the following kinds of objectives:
  - Two specific habit objectives (what, when, where, how, with whom, etc.) about the use of virtual and/or augmented reality in a tourism destination.
  - Two specific attitude objectives: Why, future behaviour, beliefs..., about the use of virtual and/or augmented reality in a tourism destination.
  - One specific consumer profile objective: describing the main characteristics of those who have some habit or who have some attitude about virtual and augmented reality.
- Finally, and based on the information found, the students should set a list of five hypotheses to be verified throughout the research in order to answer the specific objectives.

Likewise, the report should follow the APA style guidelines.

Phase II: Research Methodology

In this second phase, the students must own a personal Google account to design a survey in Google Forms. This survey needs to answer every specific objective of the research about virtual reality and augmented reality in tourist destinations. To design a questionnaire to measure the specific objectives, students should follow the below guidelines:

- Use Google Forms for the design and testing. https://docs.google.com/forms/u/0/
- Give an answer to each of the five specific objectives raised in the first phase.
- Develop a survey with its respective theoretical parts, ready to be carried out.
- Design a survey using the most adequate questions and scales for each objective.

The students must remember the question format must be homogeneous throughout the survey and comply with the parts established by the theory. It is advisable that students are familiar with the theoretical framework regarding quantitative research in order to have the information on how to design the surveys, i.e. sections and questions.

Phase III: Fieldwork

Among different sampling methods, the students may be directed towards convenience sampling to collect at least 50 valid surveys from the population that travels at least once a year from 16 years and older. The students can use different contact methods, either computer-assisted telephone interview (CATI), computer-assisted personal interview (CAPI), or via the internet, taking the advantage of the fact that the survey has been developed on an online platform in Google Forms. The students will have an outside class session to carry out field work. All the answers obtained should be entered into the survey in Google Forms, and a spreadsheet obtained containing the answers given by the respondents. This can be a new or an existing spreadsheet. Then, the students will have their database draft.

Phase IV: Data preparation and analysis

To develop the fourth phase, the students need to install PSPP software on their computers: www.gnu.org/software/pspp/.

Then, each team should get the database from Google Sheets to start the data preparation. The data preparation process includes the following steps:

- *Questionnaire checking*: Check all the surveys for completeness and interviewing quality. Verification must be undertaken that the minimum age ranges are met (over 16 years old) and that the interviewee travels at least once a year. If there are filter questions, checks should be carried out on their application.
- *Editing*: Identify illegible, incomplete, inconsistent, or ambiguous responses. Handle unsatisfactory responses by returning to the field to get better data, assigning missing values, or discarding unsatisfactory respondents.
- *Coding*: Assign a code, usually a number, to each possible response to each question. Also, specify the appropriate record and columns in which the response codes are to appear. For example, in a question about whether the interviewee is used to virtual reality, yes may be coded as 1 and no as 2, and appear in the N column for this respondent, because only one response is allowed. However, in questions that permit multiple responses, each possible response option should be assigned a separate column. It is advisable to make a codebook with the information about variables in the data set.
- *Import to a PSPP file*: Save the spreadsheet as tab-separated values (TSV) file. This kind of file is a simple text format for storing data in a tabular structure,

and each record in the table is one line of the text file. Then, open PSPP and import the file to start to analyse the database.

- *Select a data analysis strategy*: This step is based on the characteristics of the data, properties of statistical techniques, and the specific objectives and hypotheses formulated. Usually, a frequencies distribution is needed to analyse habit objectives, descriptive analysis can be applied to measure attitude objectives, and cross-tabulations and Chi-square should be used to find a consumer profile. This should be the case only if all the variables to analyse are discontinuous; ANOVA should be used if any variables implicated in the analysis are continuous. Chi-square and ANOVA are tests to confirm the association between variables.

The students must know how to identify the most appropriate kind of analysis to apply to the response to each specific objective and then interpret the statistical measures to verify the hypothesis.

Phase V: Preparation of the report

The students will prepare a report describing and showing the outcomes from each phase of the research, including conclusions and future lines of research, in relation to the results obtained. Students must examine the main objective of the research against the background and then discuss the major findings, conclusions and recommendations to destination management organisations, with the aim of helping them to make decisions about the implementation of virtual and/or augmented reality in their destination.

The report content includes:

1. Introduction of the topic to be investigated
2. Background: Secondary information and its sources
3. Specific objectives and hypotheses of the research
4. Description of the research methodology
5. Results obtained: tables, figures and/or graphs
6. Accurate interpretation and summary of results
7. Conclusions and research recommendations and suggestions.

In the last section, the results obtained are generalised, leading to the conclusions of the study stating whether the students accept or reject the proven assumptions in the analysis of results. Also, it must contain recommendations to both researchers and companies, especially to support the local tourism authorities to make decisions about the implementation of virtual and/or augmented reality in their destinations. It is advisable to include the limitations that students may identify, as well as all those proposals for improvement, in order to carry out similar research in the future.

In addition, APA style guidelines are required so that at the end of the report the students include bibliographic references consulted during their research.

References

Beck, J., Rainoldi, M., & Egger, R. (2019). Virtual reality in tourism: A state-of-the-art review. *Tourism Review*, 74(3), 586–612.

McFee, A., Mayrhofer, T., Baràtovà, A., Neuhofer, B., Rainoldi, M., & Egger, R. (2019). The effects of virtual reality on destination image formation. In J. Pesonen, & J. Neidhardt (Eds.). *Information and Communication Technologies in Tourism 2019* (pp. 107–119). Switzerland: Springer, Cham.

Yung, R., & Khoo-Lattimore, C. (2019). New realities: A systematic literature review on virtual reality and augmented reality in tourism research. *Current Issues in Tourism*, 22(17), 2056–2081.

Further reading

Barrado-Timón, D. A., & Hidalgo-Giralt, C. (2019). The historic city, its transmission and perception via augmented reality and virtual reality and the use of the past as a resource for the present: Aa new era for urban cultural heritage and tourism? *Sustainability*, 11(10), 2835.

Malhotra, N. K. (2015). *Essentials of Marketing Research: A Hands-On Orientation*. Essex: Pearson.

Case 12

CONVENT CARMEN

Rethinking the boundaries between tourism and the city

Juan Sánchez-Villar, Enrique Bigné and Luisa Andreu

Duration

The duration of the suggested four steps is flexible. It could be adapted to a three-hour session or split into two sessions of one and a half hours each. The time for reading the case study is at minimum 15 minutes, with 60 minutes for reading comprehension (questions). However, where there are oral presentations, the duration of this case study can be extended for another 60 minutes.

Learning objectives

Upon completing the case, participants will be able to:

- better understand the dynamic nature of the tourism and hotel industry
- identify the importance of authentic experiences in tourism
- practice responsible tourism and co-created value, identifying the different stakeholders in this process
- compare conventional versus contextual hotels in tourism.

Target audience

Aimed at undergraduate and postgraduate students, the case study is suitable for individual or group work. Basic knowledge on marketing is required prior to its execution. The ideal group size for the discussion is four to five students. The ideal size of the class is a maximum of 20–30 students.

Students begin by thinking about value co-creation among visitors and residents, taking into account experiences when they have travelled to a destination and/or

DOI: 10.4324/9781003182856-12

experiences in which they are acting as residents. In addition, depending on the managerial experience of the students, they can think of suggestions for both the private and public sector and to what extent these will facilitate value co-creation among destination stakeholders.

Teaching methods and equipment

Flipped classroom. This case study can be analysed individually or in groups to address the topics listed in the learning objectives. Along with the case study, references are provided to deepen the subject according to the needs of the educational level (undergraduate or postgraduate). Steps to guide teachers/instructors are outlined below.

Step 1

Initial discussion in groups. Divide students into groups (or in pairs or individually), and start with an introductory discussion session where students can have a brainstorming activity on the case topics. For instance, the teacher can ask the following questions:

1. To what extent do hotels interact with residents and visitors in your city? Is this approach sustainable?
2. Do you know of any initiative in which residents and visitors co-create value within a destination? If yes, please identify some examples. If not, do you have any suggestions to develop a hotel in which residents and visitors can interact?
3. To what extent do value co-creation and sustainability go together in developing tourism products?

Step 2

After the introduction, during which students bring examples using their knowledge/experience (working in groups, pairs or individually), teachers provide the case study of the Convent Carmen. Students should read the case individually and discuss with their groups the key ideas.

Step 3

After reading the case study, teachers continue with the provision of the same concepts/theories (value co-creation, sustainable marketing, service design, tourist experience) with the specific example of the Convent Carmen. In this step, students should demonstrate the five objectives suggested in the teaching instructions.

Step 4

Finally, as a conclusion session, hold a discussion to identify what they have learned from this specific case, and what other suggestions they can develop, etc.

Teaching instructions

The case focuses on a specific project in the city of Valencia (Spain), but the conclusions can be generalised to other cities. Students are able to:

1. explain how this authentic experience has been designed in the city of Valencia, and analyse how authentic experiences could be designed in other cities
2. understand that the project has been built from its neighbourhood scale, build relationships with multiple community agents, and know which agents should be taken into account to foster value co-creation and be socially responsible
3. identify other examples of contextual hotels in other cities, and analyse differences with traditional urban hotels
4. in the case of a traditional hotel, know what changes in the design of the service should be made to become a contextual hotel
5. justify why the case study is an example of responsible tourism.

Case

Convent Carmen is an urban hotel that reassesses the conventional boundaries between hotel and city. The project is located in the centre of Valencia (Spain), in the Barrio del Carmen[1], in a district especially sensitive to the effects of tourist activity, specifically before Covid-19, with overcrowding and a proliferation of tourism housing. The city population accounts for 795,736 inhabitants. In 2019, the number of overnight stays in Valencia exceeded five million per year, 33% of which was domestic tourism (Visit Valencia, 2021b). The Convent Carmen project aims to recover an old Carmelite convent, with a garden of almost 2,000 m^2, built in the seventeenth century. Convent Carmen is a recreation facility for city residents that also includes a hotel for tourists. It is an urban hotel with neighbourhood roots, which allows for the building of relationships between tourists and multiple agents of the community in the vibrant heart of the city (Turisme Comunitat Valenciana, 2021).

Historically, urban hotels were natural gateways for people and ideas to come together from various parts of the world. They were living, permeable places that played an important role as referents of a city. However, current hotel development dilutes that spirit of co-creation between residents, tourists and other stakeholders. Convent Carmen seeks to recover its history and expand its value proposition. The project aspires to become not only a gateway for ideas and people, but also a gateway to the identity of an entire city. To do this, it starts from the basis of creating an ecosystem through social and cultural activation that functions as a natural meeting point between those who live in the city and those who visit it for a few days. Moreover, for travellers, Convent Carmen develops an accommodation module that allows them to live from inside that ecosystem.

Three anchors for a pioneering project

Today, urban tourism is facing challenges for which traditional hospitality lacks solutions: the growing complexity of the tourist phenomenon, the continuous search for authenticity by travellers, the emergence and consolidation of accommodation formulas different from traditional ones, the homogenisation of urban environments and social protests against some forms of tourism. These issues require transversal and multidisciplinary approaches that take the concept of urban tourism further, rendering obsolete the conventional hotel's simplicity as a space where the tourist simply sleeps, has a shower and eats breakfast.

Convent Carmen was born with the willingness to transcend the conventional hotel boundaries and leave a mark, to be a real part of the city. To do this, it understands the design of its product as a more extensive process, which involves different urban agents and does not begin or end with the accommodation function. This vision of hotel development as a co-creation effort in turn draws awareness of the impact of tourism in the centre of cities and acts as a model of responsibility. Consequently, the result is a hotel project radically different from those existing in the city that, from a competitor analysis, does not stand on conventional attributes of choice. Figure 12.1 shows a summary of the three anchor points of Convent Carmen.

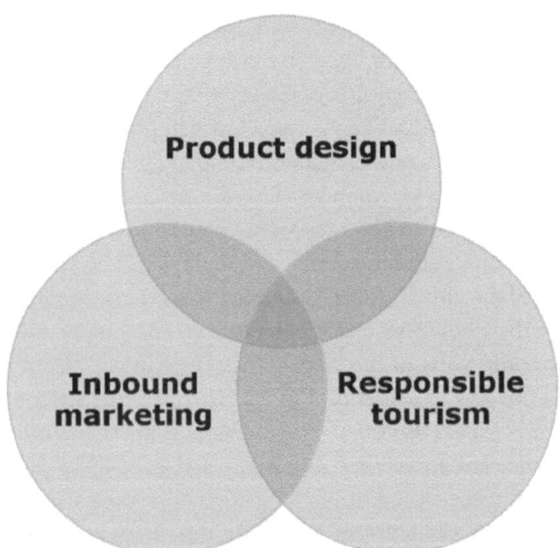

FIGURE 12.1 Convent Carmen's Anchor Points

Contextual hotel: A design beyond the existing paradigm

Convent Carmen is a hotel designed from the inside out, towards the city. This innovative approach differs from the inward design of conventional urban hotels. In order to be different from other competitors, a traditional hotel focuses only on attributes (a striking decoration, a range of services or wider gastronomic offer, or larger television screens). From this starting point, the most common attribute of urban hospitality is to act as a passive agent: feed on tourists who arrive in a city but offer little or nothing in return to the residents of it. How many hotels in the city we live in have we, as residents, actually entered? What role, what use do these establishments play for residents? The answer to both questions is the same: little to none.

The principles of traditional urban hospitality, which start from the conception of the hotel establishment as an isolated and disconnected element from the environment, are inadequate to respond to the social and cultural changes implicit in the act of travelling as a form of enrichment and personal learning. Urban tourism has undergone insightful changes and if hotels can keep up with the evolution of the city, there lies great opportunity and a certain ethical imperative for developers. In this sense, Convent Carmen is a disruptive innovation because it stops looking only at the product: it is about first understanding the environment and proposing the hotel project based on that analysis. Thus, the city's social, cultural and civil inputs that participate in the daily operation of the hotel are those that shape it. This approach is called a *contextual hotel*. It eliminates standards in terms of design and value proposition, emphasises authenticity and does not compromise scalability. Each city has features that make it unique and that diverse identity constitutes the soul of each contextual hotel. From this logic, although there are no two similar hotels, each shares the following principles:

- *Central hotel*: The hotel is born where the city emerges, where travellers merge and coexist with the locals. The meeting point does not always coincide with the geographical centre of the city.
- *Asymmetrical hotel*: The city vector determines the hotel's nature and its evolution (the antipodes of the typical theming). Each city is capable of offering different activation vectors that make it a singularity.
- *Open hotel*: A gateway to the city, but also an exit door to its identity. Capturing what traits make up a city's identity is a crucial task for the hotel developer.
- *Singular hotel*: Grows in the inhabited city, in an existing building, ideally with historical character, and from a respectful and memorable architectural intervention.
- *Co-created hotel*: The result of a process of collective construction developed together with agents of the civil life of the city.

The contextual hotel seeks to overcome the paradigm of the urban hotel as a purely extractive entity. Although generalisations are often unfair, most urban hotels are spaces alien to the heartbeat of the city in which they are located. Standardised businesses can be replicated anywhere, each of which adopt as a language theming or differentiation based on their immobilised features (larger rooms, larger beds, etc.), which leave residents indifferent, since the hotel is oriented towards the purely economic, functional and unidirectional purposes in their value proposition. In short, hotels are actually isolated entities which lack meaning for their cultural context, although of course they are completely legitimate and effective business proposals.

Faced with this paradigm, the contextual hotel adopts different starting points. In the first place, it rethinks the conventional limits of the city, opening a dialogue with its environment and transcending its tourist and economic function. To do this, it is born in places shared between residents and travellers, in squares that catalyse the local lifestyle and that become meeting points of interest. Convent Carmen is not a thematic product, and does not force anything upon the residents or tourists. It only tries to create a reflection of the life of the city, of its most genuine identity, within the hotel grounds and as part of its value proposition. Thus, such an ecosystem is itself a creative focus that reproduces (and hopefully amplifies) the quality of life of a place. By definition, urban tourism is a natural extension of the quality of life of a city that people who come from outside want to know and enjoy for a few days. Thus, the contextual hotel tries to create an ecosystem that captures how the residents of a city are – how they live and interact socially – and this creation will be what drives the tourist activity and returns part of its benefit to the context from which it is born. In summary, the distinctive features of the contextual hotel are shown in Figure 12.2.

Create community to seduce the traveller

From the point of view of its marketing strategy, Convent Carmen does not compete with the rest of the hotels in the city in the area of hotel attributes, but instead appeals directly to the traveller and their desire to live like a local for a few days. However, this is only possible if you first have an ecosystem that is a natural mirror of that local life. Following the logic of attraction-conversion-prescription

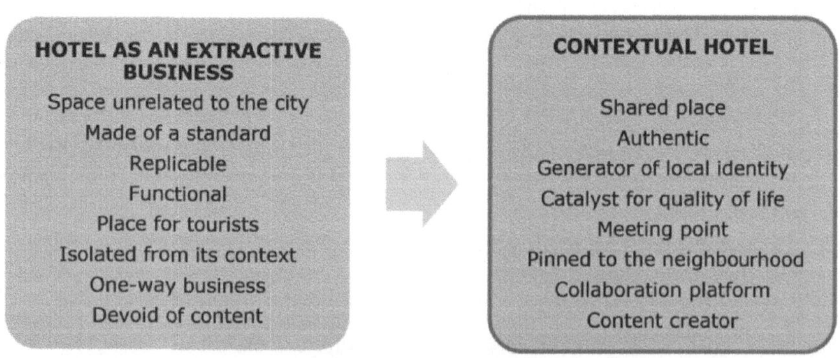

FIGURE 12.2 Distinctive Features of the Contextual Hotel

on which the ideological foundations of inbound marketing pivot, the challenge for the hotel is to attract the traveller from the prescription made by residents and other travellers of what happens daily in it. So, how to develop that place that ends up becoming something like a public square? How to first be interesting enough for residents to be a point of arrival for travellers?

Convent Carmen has achieved this balance by combining the ongoing activation of the space with the creation of an indoor gastronomic market, the first in the city of Valencia. That is, the hotel project encompasses the traditional function of accommodation together with the restaurant industry in the same place. Additionally, the project also consists of a socio-cultural centre, an economic activity completely alien to the hotel and restaurant industry that, however, is integrated organically within the organisation (see Figure 12.3). The hybridisation of all activities is a complex but necessary task for the global approach of the contextual hotel.

With regard to the gastronomic market, it is located in an open-air garden of almost 2,000 m^2. The Convent of San José and Santa Teresa, the original name of the building that houses Convent Carmen, was a Carmelite community of cloistered nuns who lived isolated from the world for almost 400 years and who had a large garden for their daily tasks. Today, this garden is the gastronomic market, open to the public with a popular and informal restaurant (see Figure 12.4). Both the food

FIGURE 12.3 Scope of Convent Carmen's Activity

FIGURE 12.4 Convent Carmen's Gastronomic Market

Source: Convent Carmen

FIGURE 12.5 Celebration of a Musical Event at Convent Carmen

Source: Convent Carmen

and drink offering and the design of the garden promote sharing between visitor and resident. The pieces of furniture combine different typologies and are designed to encourage social interaction, for the appearance of spontaneous exchanges of a personal nature between the people who meet in the place.

On the other hand, the socio-cultural centre constitutes the true cornerstone of the place's activation strategy. In this sense, the convent has a desacralised church that, with a capacity of 300 people, allows the celebration of all kinds of social and cultural events. The garden includes a stage that complements the social activities together with the gastronomic market (see Figure 12.5). Its arena setting coexists in a natural way, providing both hospitality and cultural functions. The hotel is the container for all this vibrant activity, which changes every day and reflects the lifestyle of an entire city.

The task of socio-cultural programming activities is designed and curated by its own staff, but is nourished by alliances with the city. During the first 17 months of operation, Convent Carmen was able to host almost 350 events configured around a series of thematic axes (live music, debates of ideas, performing arts, workshops, literary events, cinema, dance, gastronomy, well-being, etc.) and established a network of collaborators of more than 100 agents of the city. From the marketing angle, the creation of specific thematic foci is not accidental: it allows the project to expand its scope and correctly segment its target audiences according to the day, the content and the time slot. This gives absolute transversality to the value proposition and eliminates the possibility of appropriation of the project by a specific urban

FIGURE 12.6 Sample of a Monthly Sociocultural Programme at Convent Carmen

Source: Convent Carmen

group. Tolerance, coexistence and the absence of prejudice are distinctive features of the programming (see Figure 12.6).

Acting as a platform for what happens in the city and putting the space at the service of local human talent lays the foundation for the creation of a cohesive community that makes Convent Carmen a permanent point of attraction. During its first 17 months of operation, the space received more than 400,000 visitors, generated a remarkable network of virtual followers (18.8k on Facebook and 21.3k on Instagram) and had more than 300 appearances in various media. Such notoriety gives the project the potential for tourist attraction regardless of the attractiveness offered by the city itself. The relationship is bidirectional: the project is interested in the small shining, that initiatives that have hardly left the rehearsal room have the opportunity to be discovered by a wider audience, which thus promotes the surprise and viralisation of the spontaneous.

Moreover, the public is interested in a place that offers varied, accessible content with a low barrier for entry, and that also allows informal socialisation in its gastronomic garden. To the extent that this balance is maintained, the daily beat of the place is a mirror of the daily beat of the city, a lifestyle ideal that will seduce travellers who want to be part of this ecosystem. Only after achieving all these balances does the existence of a conventional accommodation function, and the concept of hotel as it is usually understood, make sense. That is why Convent Carmen, despite being an active agent of the city, does not compete in programming against other socio-cultural centres (museums, theatres, concert halls, etc.); rather, it is their ally, but it does pin itself against the rest of the hotels in the city. The results seen in that area demonstrate a positive balance for the business model and for the city.

Specifically, through cultural action, Convent Carmen seeks to offer visibility to agents operating on the regulated margins of the sector and facilitate access to culture for the general public. While a large number of the events that are part of the programming calendar are freely accessible, the spirit of implementation is fully professional. Convent Carmen tries to understand the operating model of each activation and balances the interaction with the collaborating agent, rewarding it directly or indirectly so that it is both economically feasible and provides interesting content. All the resources of this programming task are private, raised as part of a broader business model, the hotelier, and have the direct contribution of almost a dozen sponsors, which grants flexibility to decision making and allows total autonomy over the result.

As for social action, Convent Carmen has been weaving a dense network of relationships with various civil agents of the city that has allowed results ranging from the recovery of the historical memory of the convent in which the hotel is located, the labour integration of people at risk of exclusion, collaboration with the soup kitchen of the NGO Messengers of Peace or the empowerment of the Senegalese female community of a marginal neighbourhood in Valencia. The most regulated lines of collaboration have also resulted in agreements with entities such as the Red Cross, Tramundi CH, Vicente Ferrer Foundation, Pequeño Deseo Foundation or Oceanogràfic Foundation, among others. All this connects the tourism project with an expanded function of responsibility that puts its focus on local development and culture, which is detailed in the following section.

Responsible tourism: transcending the economic function

Given its pioneering character, Convent Carmen is a sequential project in which each stage settles the conclusions of the previous phase. As the phases of development of the concept and selection of the space moved from theory to practice, the project created the ideal framework for the implementation of a key phase, the activation of space, as established in Figure 12.7.

The stage of activation of the space has similarities with an exercise of placemaking insofar as it is oriented to a local audience, recovers a place for public use, develops collaboration and produces effects that go beyond the limits of the business. Convent Carmen has chosen cultural and social action as reflections of the identity of Valencia, the city in which it is located, but other vectors of activation could also be raised. In any case, conceiving a hotel based on its relationship with the city or managing a private space subordinated to public interests, such as free access to culture or social integration, are just a sample of Convent Carmen's

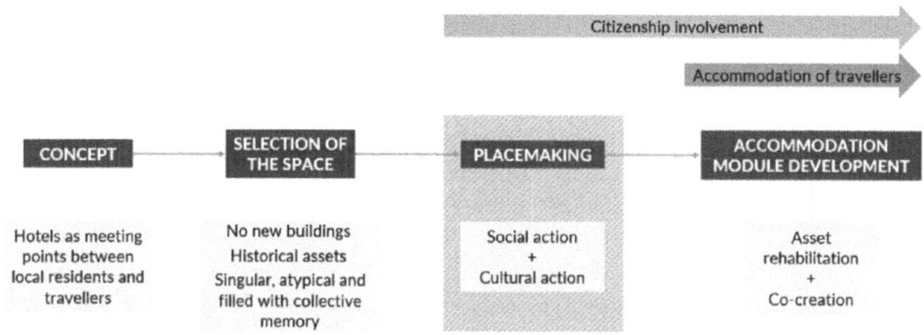

FIGURE 12.7 Convent Carmen's Longitudinal Process of Development

intentions to transcend its economic function. In addition, the hotel's development is an exercise in the recovery of the heritage of a disused and remarkably degraded seventeenth-century religious site. The project thus stands as a model that understands sustainability from its various edges (economic, environmental and social) and that seeks the creation of a better city through a more conscious, responsible and inclusive form of tourism.

Note

1 The 'Barrio del Carmen' (El Carmen Neighbourhood) is an emblematic neighbourhood found in the city's old quarter. It is the authentic historical centre of Valencia (Visit Valencia, 2021a).

References

Turisme ComunitatValenciana (2021). Convent Carmen, a New Cultural Proposal in València. https://www.comunitatvalenciana.com/en/be-inspired/convent-carmen-a-new-cultural-proposal-in-valencia

Visit Valencia (2021a). El Carmen Neighbourhood. www.visitvalencia.com/en/what-to-see-valencia/historical-centre/del-carmen-neighborhood.

Visit Valencia (2021b). Tourism Statistics. https://fundacion.visitvalencia.com/sites/default/files/media/downloadable-file/files/folleto-estadisticas-2019.pdf

Further reading

Amaral, P. (2012). Entrevista Phillip Kotler [Interview with Philip Kotler]. YouTube, May 28. www.youtube.com/watch?v=TpR0Ij6p2Xw.

Beck, D., & Storopoli, J. (2021). Cities through the lens of stakeholder theory: A literature review. *Cities*, 118, 103377.

Convent Carmen / Francesc Rifé Studio (2018). Convent Carmen / Francesc Rifé Studio. ArchDaily, October 23. www.archdaily.com/904495/convent-carmen-francesc-rife-studio>

Kotler, P., Kartajaya, H., & Setiawan, I. (2021). *Marketing 5.0: Technology for Humaniity*. Hoboken, NJ: Wiley.

Kozak, M., Rita, P., & Bigné, E. (2018). New frontiers in tourism: Destinations, resources, and managerial perspectives. *European Journal of Management and Business Economics*, 27(1), 2–5.

Timur, S., & Getz, D. (2008). A network perspective on managing stakeholders for sustainable urban tourism. *International Journal of Contemporary Hospitality Management*, 20(4), 445–461. https://doi.org/10.1108/09596110810873543

Turisme Comunitat Valenciana (2021). Convent Carmen, a New Cultural Proposal in València. www.comunitatvalenciana.com/en/be-inspired/convent-carmen-a-new-cultural-proposal-in-valencia

van Riel, A. C., Andreassen, T. W., Lervik-Olsen, L., Zhang, L., Mithas, S., & Heinonen, K. (2021). A customer-centric five actor model for sustainability and service innovation. *Journal of Business Research*, 136, 389–401.

Case 13

HOTELS AND ONLINE TRAVEL AGENCIES

A partnership made in heaven?

Peter O'Connor

Duration

Depending on whether participants prepare the case in advance, or in buzz-groups in class, two different durations are possible, oulined in Table 13.1.

Learning objectives

Upon completing the case, participants will be able to:

- help participants understand the positives and negatives of online travel agents as distribution partners
- highlight the costs and difficulties of driving online bookings directly
- qualitatively and quantitatively evaluate alternative ways in which to drive online bookings
- explore current issues in distribution for hotels (and other tourism businesses).

Target audience

The case is designed to support discussions on the value of online travel agencies within the distribution strategy of hotels. As such, it can be used in classes dealing with tourism or hospitality marketing, revenue management, online marketing or broader marketing courses addressing on the role of intermediaries. While a broad familiarity with the tourism industry would be useful, all information needed to work on and understand the case is included within the case text and support. The material is pitched at the bachelor's and master's level, but the topicality and controversial nature of the topic means that it particularly stimulates (sometime heated) debate among industry professionals.

DOI: 10.4324/9781003182856-13

TABLE 13.1 Alternative Durations

	Prepared in Advance	*In Class*
Introduction	5 mins	5 mins
Q1: Critically evaluate the hotel's current distribution strategy.	10 mins	10 mins
Q2: Should hotels work with online travel agents? What are the advantages and disadvantages of doing so?	15 mins	Buzz group: 15 mins Debrief: 15 mins
Q3: Are the two ideas identified by the management to decrease the hotel's dependence on online travel agents likely to be successful? Critically evaluate the two options and explain their implications for the hotel's operations and profitability.	20 mins	Buzz group: 15 mins Debrief: 20 mins
Q4: (Optional) What other suggestions do you have to improve performance?	5 mins	5 mins
Summary	5 mins	5 mins
Total	60 mins	90 mins

Teaching methods and equipment

Designed to be used to stimulate thought and discussion, both between participant groups working on preparing the case, and in the classroom debrief discussion itself, the case can be used at two levels.

- At a more basic level, the macro level issues, including the effectiveness of the Majestic Hotel's current distribution strategy, the advantages and challenges working with online travel agencies in general, and the likely effect/success of the alternative solutions proposed by the Dubois family, can be explored.
- However, given the quantitive data included in the case, a more detailed analysis can also be performed, in particular financially evaluating the option of using online marketing techniques, such as paid search engine marketing, to drive direct bookings.

Both approaches should lead the learner to the same conclusion and clarify the value added from working with intermediaries such as the online travel agencies.

Teaching instructions

Participants should read the case in advance, familiarising themselves with the hotel's current situation and thinking about the challenge introduced at the end of the text. The suggested questions (shown below) can be assigned in advance, or can be introduced during the class discussion.

To stimulate discussion, four broad sequential questions are proposed:

1. Critically evaluate the hotel's current distribution strategy.
2. Should hotels work with online travel agents? What are the advantages and disadvantages of doing so?
3. Are the two ideas identified by the management to decrease the hotel's dependence on online travel agents likely to be successful? Critically evaluate the two options and explain their implications for the hotel's operations and profitability.
4. What other suggestions do you have to improve performance?

Suggested answers to each of these questions are shown below.

Introduction

The case is designed to highlight and stimulate discussion on the role and importance of online travel agencies (OTAs), as well as highlight many of the challenges facing hotels today as consumers increasingly shop and book through the online environment.

Instructors should start by pointing out both the high fixed-cost nature of the hotel business (see the Income Statement for the Majestic Hotel in Table 13.6), where the majority of the costs must be paid irrespective of the number of rooms sold and also the highly perishable nature of the hotel product (an unsold hotel room cannot be stored and sold at a later date), which combined make the sale of as many rooms as possible every night important. As a result, distribution, or to put it simply, facilitating the sale of hotel rooms, has become one of the key success factors for hotels.

While the growth of the web as a consumer shopping and booking channel has in one way made it easier for hotels to reach out to customers and sell their rooms, independent hotels (which make up the majority of properties in Europe) often lack the appropriate expertise, technology-based systems and marketing budget to be able to exploit the opportunity of online channels. Many instead rely on online travel agents such as Booking.com, Expedia Inc. (which includes brand names such as Expedia.com and Hotels.com), eDreams and Lastminute.com, to service online demand.

It is worth pointing out in the introduction that many hotels have a love-hate relationship with the online travel agents. While they appreciate the business driven by such systems, they resent the seemingly high commissions that must be paid in return for these bookings. This relationship has become increasingly confrontational as the online travel agents have gained market share and drive a higher proportion of bookings, with some hotels accusing them of charging high commissions on bookings that they would have gotten anyway.

Some hotels think they can drive online bookings directly, reducing commissions and overall costs. However, whether this would occur or not is questionable as, contrary to the thinking of some hoteliers, direct bookings do not just arrive on your doorstep unprompted, but require concerted marketing effort and expenditure to make them happen.

Q1 Critically evaluate the hotel's current distribution strategy

To place the case in context, the best place to start is by looking critically at the hotel's current performance as regards distribution. In a plenary session, instructors might start by asking participants their opinion as to whether Nicole is doing a good job in managing the distribution of the hotel (and why)? A good follow-on question to stimulate further discussion would be to ask what challenges she currently faces and how she could further improve performance.

An overview of the key points that should arise in this discussion is provided below.

On the positive side, what they are currently doing seems to be working. When we look at Table 13.2, we can see that, with relatively little attention and effort, the hotel is currently outperforming its competitive set (similar hotels in the area) in terms of both occupancy and average daily rate. Considering that many of these competing properties would be members of various hotel chains, and thus would get a boost from flying the flag of the brand, as well as profit from its expertise and expenditure on advertising, sales, marketing and distribution, this is quite an achievement.

However, this success at driving high quality business comes at a considerable cost. Well over one-third of the hotel's reservations come from an online travel agent and thus necessitate the payment of a commission of approximately 17% of room revenue. Over the past number of years, this proportion has grown steadily, to

TABLE 13.2 Comparative Market Performance, Majestic Hotel, 2019

2019 Month	*Occupancy*	*Average Daily Rate*	*Competive Set Occupancy*	*Competive Set ADR*
January	56,20%	€106	55,50%	€105
February	54,60%	€107	55,90%	€105
March	61,50%	€106	64,00%	€105
April	66,70%	€104	69,70%	€103
May	68,90%	€106	68,30%	€107
June	80,30%	€125	81,50%	€123
July	78,70%	€106	78,00%	€113
August	78,30%	€111	77,70%	€110
September	79,20%	€123	78,00%	€120
October	69,10%	€103	70,50%	€107
November	62,10%	€99	62,00%	€101
December	55,10%	€109	54,10%	€107

Competetive Set Source: In Extenso Tourism, Culture & Hospitality (www.inextenso-tch.com)

the detriment of both direct and tour operator bookings, and now represents nearly 38% of bookings. This makes the hotel highly reliant on the online travel agents and perhaps suggests that they should diversify their sources of business.

Efforts to drive bookings through the hotel's own direct channels (be they off-line or online) seem increasingly ineffective. While more and more customers are booking through online channels in general, the proportion of bookings flowing through the hotel's own website is declining, despite the rate being cheaper (and better value since it also includes breakfast) than both online travel agents and off-line direct channels.

A variety of issues place a question mark over the hotel's commitment to trying to distribute directly. The website clearly needs refreshing and the booking engine needs to be replaced with a more modern, multilingual tool. Marketing spend, at 2.7% of gross revenues, is low, and Nicole, while competent, only spends a portion of her time on distribution and online marketing issues. Right now, it seems that the hotel is making little to no effort to drive bookings directly, instead preferring to sit back and let online travel agents do the work for them. But as the proportion of bookings flowing through online travel agents creeps up, this become an increasingly costly, and perhaps risky, strategy.

While working closely with online travel agents has been beneficial until now, the hotel needs to carefully evaluate (both financially and strategically) this approach for the future. Perhaps the best place to start is by examining the positives and negatives of working with online travel agents in general.

Q2 Should hotels work with online travel agents? What are the advantages and disadvantages of doing so?

While initially welcomed by suppliers as an additional way to reach out to the marketplace and sell rooms, sentiment towards online travel agents turned less positive as they become more successful and captured increased market share. Some hotels grumble that online travel agents' extensive marketing efforts steal business away from the hotel, or force them to pay high commissions for bookings that they otherwise would have gotten for free.

While there may be some truth to these accusations, often they are based on an incomplete or biased understanding of what online travel agents actually do and how they add value. Therefore, an objective evaluation of the advantages and disadvantages of working with online travel agents would be useful to frame the discussion.

Participants can be assigned into buzz groups of three to four people to perform this analysis (10 mins), or the question can be posed in advance and participants asked to individually prepare their arguments.

In either case the instructor should debrief the exercise, highlighting the positives and negatives of the hotel/OTA relationship on the board or flipchart.

Several of the key points usually raised are outlined below.

Advantages

- *Visibility*: Working with the right online travel agents gives the hotel visibility in markets/sectors that they would find difficult to address individually. For example, partnering with an online travel agent means that your property is immediately available for sale in foreign markets in which the online travel agent operates, without any further expenditure on marketing, translation costs or credit card risk. As such online travel agents can be a good source of incremental bookings – business that you would not have received had you not been listed on the system. For the Majestic Hotel, it's worth noting that the majority of France's major source markets are other European countries in which online travel agents are widely used by consumers to shop and book leisure travel, so being listed on the right portfolio of online travel agents likely gives it very valuable exposure.
- *Pay-per-performance*: Distributing through online travel agents is generally commission based and thus 100% pay-per-performance, with no reservations meaning no costs. This contrasts sharply with other forms of sales and marketing expenditure, where spending must be done up front with little, if any, guarantee that it will result in any supplemental business.
- *The Billboard Effect*: In addition to indirect bookings generated through the online travel agent channels, there is evidence that simply being listed on online travel agent sites results in increased (online and offline) direct bookings. Even though consumers search on online travel agent sites, many subsequently visit the hotel's direct website and/or call the hotel directly once they have found something that potentially meets their needs. Assuming that the hotel's offer is competitive (in terms of both product and price), this often result in additional, free from commission, bookings through direct channels.
- *Keeping up with technology*: While hotels are not well known as early adopters of technology, travellers continue to push the boundaries in terms of how they search for and book travel. For example, many now search and book through mobile devices rather than over the web. Similarly, some are experimenting with voice-, messaging- and even virtual reality, -based channels. Rather than continually playing catch-up, hotels can (initially at least) delegate this role to the online travel agents, many of whom have technology at their core and, working on behalf of multiple suppliers, can leverage economies of scale to exploit new and developing channels.
- *Corporate Intelligence*: Servicing multiple suppliers and travel segments, online travel agents consolidate vast amounts of real time data that can be used to generate powerful market-level intelligence. Increasingly online travel agents are sharing these insights with their supplier partners, serving as a valuable input into forecasting, capacity planning and revenue management, and helping them to manage their distribution better.

Disadvantages

- *Perceived cost*: OTA bookings typically necessitate the payment of a commission, typically in the range of 15% to 17% for independent properties. Since this needs to be paid every month, such commissions are highly visible, with someone needing to write a cheque to each OTA each and every month. This contrasts with, for example, tour operators, which, even though the net revenue is lower and costs are higher since guests get a free breakfast, is not perceived as expensive as its cost is hidden since it is a discount rather than a commission. A key question to be answered (see later) is whether OTA commission costs are actually high, particularly compared to the cost of driving equivalent bookings through direct channels.
- *Loss of control*: A power imbalance undoubtedly exists between the large online travel agents and their (in particular independent) hotel suppliers. Over the years, this has allowed online travel agents to dictate terms and working practices. For example, in the past some online travel agents demanded both best available rate and last room availability in order to be listed on their systems. However, as this limited hotel ability to compete, such rate and inventory parity clauses have been formally outlawed in many countries, although evidence remains that online travel agents still penalise hotels that offer cheaper rates on alternative channels by disadvantaging them in their display algorithm.
- *Channel cannibalisation*: With limited marketing budgets, many hotel properties struggle to gain and maintain visibility in the online environment, particularly when multiple online travel agents are selling, and thus marketing and promoting, their property online. Coupled with the superior merchandising expertise employed by most online travel agents to convert prospects into sales, this undoubtedly results in some business that might have booked direct flowing through indirect channels. Such channel cannibalisation is particularly likely to occur when hotels do not do a good job of managing the competitiveness of the price and product offerings on direct channels.
- *Over-dependence*: Given their ease, efficiency and cost effectiveness compared with trying to drive business directly, it's easy for hotels to get lazy and just delegate online selling to OTA partners. However, being overly dependent on any one source of business can be problematic. If that channel were to fail, or disadvantageously change its business practices, this would have severe implications for the future of the hotel. Even though it may require more work, maintaining a balanced portfolio of distribution channels is essential to mitigate risk.

At the end of the discussion, the instructor should summarise and bring together the points made, highlighting that while distributing through online travel agents has many advantages, it also comes at a (strategic and financial) cost, and hotels need to carefully evaluate whether, when and how they should use online travel agents as distribution partners. In particular the link should be made with the Majestic Hotel, which is receiving increased proportions of its business from online travel agent channels, with a corresponding increase in commissions paid, at a time when the hotel wants to increase profitability.

The Majestic Hotel's management have suggested two alternatives ways of addressing its OTA challenge – closing out/putting less favourable rates on OTA channels and/or trying to drive more direct business by investing in paid search engine marketing. Participants should then be asked to evaluate these options and compare them with the option of maintaining the status quo with the online travel agents.

Q3: Are the two ideas identified by the management to decrease the hotel's dependence on online travel agents likely to be successful? Critically evaluate the two options and explain their implications for the hotel's operations and profitability.

Once again, these two questions can be pre-assigned for preparation or discussed in buzz groups. In both cases participants' suggestions should be debriefed in a plenary session where the following key points should be emphasised.

1. *Jean Francois's idea of closing out/putting less favourable rates on the online travel agents.* (5 minute debrief)
   The relationship between hotels and online travel agents can best be described as symbiotic. While hotels need online travel agents to sell their rooms, online travel agents also need hotels to provide them with rooms, at times and prices that they can sell, to generate revenue. If either party does not fulfil their end of the bargain, the relationship breaks down and no one wins.

   If the hotel were to close out their OTA channels during busy periods, this would leave the OTA with no rooms to sell at precisely the time at which demand is high. As a result, it would not be able to satisfy its customers and would not make money. Similarly, if the hotel were to charge higher room rates on OTA channels, thanks to the aforementioned Billboard Effect, customers might shop on the OTA but book on the direct website as a result of its lower price, resulting in no revenue for the OTA. In such cases, the hotel is not playing the game and giving the OTA a reasonable chance of being able to sell the hotel, so why should the OTA feature that property in what it proposes to potential customers?

   Hotels that play games like this are likely to find themselves deemphasised on OTA search and destination listings. If potential customers directly search for them, they will be displayed and can be booked, but online travel agents are unlikely to feature them in more generic searches. So, in effect, this would have the opposite effect to the one desired, cannibalising bookings from customer who already know the property but not generating any incremental business.

   Thus, while Jean Francois's idea is likely to achieve its objective and reduce the proportion of OTA bookings and thus distribution cost, it is in fact a false economy, with overall sales revenue decreasing without something to fill the void left by the lost bookings, leaving the hotel in a much worse place financially. Most hotels, including the Majestic Hotel, run at about 70% occupancy, meaning that three out of ten rooms are vacant most nights. Instead of trying to find ways to decrease OTA bookings, the focus should be on finding ways to drive more business overall. As such, Nicole's idea of trying to drive more direct bookings through paid search may be very relevant.

2. *Nicole's idea of driving more direct business by investing in paid search engine marketing.* (15 minute debrief)

From the case text, it's clear that one of the Majestic Hotel's key challenges is its lack of visibility in the online environment. Even if a potential customer is aware of, and wants to book, the Majestic Hotel, it is difficult to find the direct website among the large number of search listings from online travel agents trying to drive customers to book through their system.

One way to increase visibility is to engage is paid search engine marketing, positioning the hotel in front of customers searching for specific keywords on the major search engines. These keywords may be generic (e.g. 'hotel Paris', 'cheap hotel Paris') or very specific (e.g. 'Majestic Hotel Paris'). Whether the hotel's listings are displayed depends on how much they are prepared to pay for each resulting click, with Table 13.3 showing the minimum bid currently needed to be displayed at the top of the first page of search results when someone searches for the specified keyword on Google.

However, each click does not result in a room sale. Instead it takes the potential customer to a page on the hotel's direct website, where they must subsequently be convinced to buy. In general, with the high traffic coming from paid search channels, hotel websites have a conversion rate of about 3%, meaning that out of every 100 prospects that click on the advert on Google, only three will buy. However, from the case, we have seen that the Majestic Hotel's website and booking engine are already dated, so the conversion rate is likely to be lower than the industry average. For that reason, a conversion

TABLE 13.3 Cost Calculation – Specific Keyword

Keyword: Majestic Hotel Paris	
Cost-per-click	€0,52
Conversion rate	2%
Customer acquisition cost	€34,67
Average daily rate	€100,00
Average length of stay	1,7
Revenue	€170,00
Percentage cost	20%

TABLE 13.4 Sample Cost Calculation – Generic Keyword

Keyword: Cheap Hotels in Paris	
Cost-per-click	€0,47
Conversion rate	1%
Customer acquisition cost	€47,00
Average daily rate	€100,00
Average length of stay	1,7
Revenue	€170,00
Percentage cost	28%

rate of 2% has been used in the example below. Dividing the cost-per-click (CPC) (€0.52) by the conversion rate (2%) gives us a cost per reservation of €34.67. Assuming an average daily rate of €100 and an average length of stay of 1.7 nights, this means that the customer acquisition cost in percentage terms is 20%, or higher than the 17% commission that would have been paid to an OTA for this reservation.

Even though the CPC is usually lower, the resulting cost is likely higher when potential customers type in generic keywords (Table 13.4). If the hotel were to bid, for example, on the phrase 'cheap hotels in Paris', the cost per click would be €0.47. However potential customers typing in such a phrase are less likely to know (and thus purchase) the Majestic Hotel. In addition, they are still likely in the shopping phase of their decision-making process, simply searching for options, and therefore more difficult to convert into actual sales. Assuming a generous conversion rate of 1%, this would result in a cost per reservation of €47.00, or 28% of the projected room revenue.

It's worth pointing out that the above calculations refer just to the pure transaction costs. If the hotel were to pursue this approach, they would have to devote considerable time and effort to managing their online marketing, implying an additional labour cost. In addition, it would likely have to redevelop its website and implement a new booking engine, implying further investment. And, unlike OTA distribution, which is pay-per-performance, trying to drive more direct bookings involves more risk. Hundreds of potential customers could click on the hotel's adverts on Google (resulting in substantial costs), but unless the direct website does a good job of converting this traffic into actual bookings, the hotel will be left with all of the costs and no additional revenue.

Q4 (Optional) What other suggestions do you have to improve performance?

This is obviously a very broad question designed to stimulate discussion on how to improve distribution performance. Common operational level suggestions include devoting more time and resources to managing distribution as currently only a portion, probably a minor portion, of Nicole's time is devoted to this vital function. Another common suggestion is to implement a technology-based Channel Management system, which would help the hotel to more consistently implement its pricing strategy, be that rate parity or using price as a competitive lever to convert more direct bookings.

On a more strategic level, the key question that arises is whether the hotel should sign up with one of the major hotel brands such as Accor, Choice Hotels International, Hilton or Marriott. Where this issue does not come up naturally in discussions, it can perhaps be promoted by the instructor.

In fact, becoming a member of a major hotel brand could address many of the challenges currently being faced by the Majestic Hotel. Hotel brands typically invest heavily in distributing their properties, spending large amounts on offline and online marketing and maintaining loyalty programmes to get customers to book directly. While most also work with online travel agents, they can also negotiate centrally for all properties in the brand to get substantially lower commissions (e.g. 12%), thus driving down this cost for their members. And, lastly, simply flying the hotel chains flag generally results in higher occupancy and average rate thanks to the increased visibility and perceived quality of the brand.

However, these advantages come at a cost. Hotels that are part of a brand need to conform with brand standards (which range from what kind of cereal they can serve at breakfast all the way up to requiring periodic costly renovations of rooms and public spaces), limiting flexibility in what they can do and necessitating substantial capital and operational costs. Hotel brands also charge substantial fees for their services. When royalty fees, loyalty programme fees, reservation fees and marketing fees are totalled, they can often collectively reach 10% of overall room revenue. And it is worth pointing out that, in contrast to OTA fees, which are based on business delivered, brand fees are charged on total room revenue. Calculating at what point the increase in business from being part of a hotel brand balances out these increased costs and results in increased profitability can be an interesting exercise.

Summary

To wrap the case discussion and bring everything together, the instructor should bring everything together by summarising prior discussions. Key issues that should be stressed include:

- Online travel agents may be perceived as expensive, but that cost is relative. Independent hotels, lacking the time, expertise and systems to be able to engage in online marketing in a meaningful way, often struggle to drive direct bookings in a more cost-effective manner. In today's highly competitive environment, there is no such thing as a free lunch when it comes to distribution.
- While direct bookings are often perceived as cheaper than OTA bookings, the real cost of driving that reservation must be taken into consideration. As we

have seen with the paid search examples in the case, often just the direct transaction costs alone are higher than what would be paid to an OTA when you actually crunch the numbers.

- Working with online travel agents must be a win-win situation for both parties. If hotels want online travel agents to generate incremental bookings, they must provide them with rates and availability that are competitive.
- Most importantly, with most hotels running at 70% occupancy, the objective of their distribution strategy should not be to reduce OTA bookings, but to find ways to increase profitability. With this in mind, rather than trying to replace OTA bookings, they should supplement them with their direct marketing efforts. While this will result in higher distribution costs, the increased number of rooms sold will more than compensate for the increased cost, resulting in higher profits. Only in the very rare case where the hotel is full should it be necessary to prioritise one channel over the other.

Case

The Majestic Hotel is an independent three-star property located in the touristy Left Bank area of Paris, France. The property, comprising 62 bedrooms, as well as a restaurant and bar, has been in the Dubois family for three generations, serving both French and international clients visiting Paris primarily for leisure purposes. The entire hotel was extensively refurbished in 2010, with the owner borrowing significantly to pay for the building works. This has placed increased financial strain on the business as both the interest and a portion of the capital must be repaid each month. As a result, more than ever before, the hotel's management is focusing on maximising both revenues and profitability.

The management team is led by Jean-Francois, the oldest of the three brothers and the patriarch of the family, who serves as managing director of the company and general manager of the hotel. Jean-Francois attended a leading Swiss hotel school in the 1970s, and while his training was primarily focused on guest service, over the past four decades he has developed a strong head for business. As a result, he is becoming increasingly concerned about the relatively large, and seemingly ever increasing, amounts of commission being paid to online travel agents each month in return for bookings.

As more and more consumers search for, and book, their travel online, the percentage of business flowing through one or more of the major online travel agents has been gradually increasing (see Table 13.5). The Majestic Hotel works with the three biggest players (Booking.com, Expedia and Ctrip.com), as well as with a select number of smaller, more niche players specialised in the Parisian market, and these channels now deliver nearly one third of the total number of rooms booked. And with each and every booking resulting in a commission (typically in the range of 15% to 17% of the resulting room revenue), these monthly payments are having a significant effect on the profitability of the business (see Table 13.6). Right now, they represent the second largest expense (after labour) for the hotel, which has brought them to the attention of Jean Francois.

TABLE 13.5 Channel Breakdown for the Majestic Hotel, 2015–19

	2015	*2016*	*2017*	*2018*	*2019*	*Net ADR*	
Direct (phone, fax, walk-in)	61%	60%	58%	55%	53%	€ 109,00	
Tour operator	8%	7%	6%	6%	5%	€ 73,00	⋆
Own direct website	6%	6%	5%	4%	3%	€ 102,30	⋆
OTAs	25%	27%	31%	35%	38%	€ 86,32	

Net ADR (average daily rate) represents the amount the hotel earns after commissions and discounts.
⋆ Includes breakfast

TABLE 13.6 Income Statement, Majestic Hotel, 2019

	Revenue	*Expenses*	*Income*	*Percentage*
Rooms department				
Rooms revenue	€1.672.463			100,00%
Expenses				
Labour cost		€162.229		9,70%
Housekeeping		€91.985		5,50%
Commission		€113.462		6,78%
Total expenses		€367.676		
Department income			€1.304.787	78,02%
F&B department				
F&B revenue	€675.250			100,00%
F&B expenses		€523.319		77,50%
Department income			€151.931	22,50%
Other department				
Other revenue	€58.536			100,00%
Other expenses		€51.570		88,10%
Department income			€6.966	11,90%
Total operationg department revenue	€2.406.249			100,00%
Total operationg department income			€1.463.684	60,83%
Undistributed expenses				
A&G payroll		€98.656		4,10%
Bad debt expense		€6.016		0,25%
Advertising and promotion		€64.969		2,70%
Maintenance		€62.081		2,58%
Utilities		€66.894		2,78%
Total undistributed expenses		€298.616		
Gross operating profit			€1.165.068	48,42%
Occupation costs				
Depreciation		€336.875		14,00%
Property taxes and insurance		€24.303		1,01%
Interest – long term		€508.922		21,15%
Lease amortisation		€6.737		0,28%
Total occupation costs		€876.837		
Profit before other items			€288.231	
Interest – short term		€3.056		0,13%
Income tax		€75.075		3,12%
Net income (loss)			€210.100	8,73%

The hotel does have its own website (www.themajesticparis.com). Developed in 2015 at a cost of Euro 30,000, this features information about the hotel in English, French and Chinese, as well as a bookings engine (English only) that integrates with the hotel's Property Management System to give access to live availability and rates. While initially this worked well, more recently booking volumes have fallen, despite the vast increase in the number of people using the web to shop and book their accommodation. Informal feedback from customers indicates that not only is it difficult to find the hotel's website among the forest of offerings from online travel agents and their affiliates, but also that the site, and in particular its booking engine, has become dated and clunky, discouraging potential customers from booking.

Distribution for the property is managed by Nicole, the niece of Jean Francois, who also works as the head receptionist. Having graduated from a leading hotel school in France, she worked for some years in the reservations department of one of the major European hotel chains, and thus she has a good knowledge of the hotel distribution environment, as well as some basic knowledge of revenue management. To avoid confusing its customers, the property tries to maintain a consistent rate strategy across all of the channels through which it is distributed, although in practice this is not always the case due to the time consuming and labour-intensive process of manually updating information on the intranets of the various OTA channels. Due to the time pressures of day-to-day hotel operations, particularly during the busy summer period, sometimes these updates tend to not get done, resulting in different rates been shown on different points-of-sale.

The fact that rates sometimes get out of sync is also problematic from the view of exploiting the developing metasearch concept. While this was originally popularised by dedicated travel metasearch sites such as HotelsCombined.com, Kayak and Trivago, similar price comparison facilities are increasingly being built into other sites, including the widely consulted TripAdvisor network and even the Google search engine. As a result, any inconsistencies in price for any specific date are immediately visible to the customer, resulting in, at best, the customer booking on the lowest price channel or, at worst, confusion and the customer selecting an alternative property. Right now, as an independent property, the hotel is not directly listed on the metasearch channels, and customers only find OTA- and tour operator-based listings for the hotel when they search on such systems, but their growing popularity among consumers means that Nicole is considering getting listed directly.

Thus, although consistently delivering a significant number of bookings for the property, OTA-based distribution, and its effect on profitability, has become a key concern for Jean Francois and the other family members involved in running the business. Nicole would like to try to drive some more direct business by engaging in some online marketing, and in particular paid search engine marketing, to boost the visibility of the hotel's website. She has investigated the cost of bidding on various Google AdWords (see Table 13.7) but is unsure how to translate these costs-per-click fees into their equivalent cost per reservation.

TABLE 13.7 Cost-Per-Clicks for Google AdWords, 2021

Keywords	*Monthly searches*	*Competition*	*Cost-per-click*
Majestic Hotel Paris	**1000**	**Low**	**€ 0,52**
Hotel Paris	**90400**	**Medium**	**€ 0,85**
Cheap hotels in Paris	**5400**	**Medium**	**€ 0,47**
3 star hotel in Paris	**720**	**Low**	**€ 0,29**
Hotel Paris left bank	**210**	**Low**	**€ 1,12**
Best small hotels in Paris	**50**	**Medium**	**€ 0,84**
Charming hotel paris	**50**	**Medium**	**€ 1,27**

However, Jean Francois is concerned about how this might affect the hotel's marketing spend and, since he doesn't really know much about online marketing, is quite resistant to the idea. And, in any case, focusing on this strategy would involve having to redevelop the hotel's website, and potentially implementing a new bookings engine, to address some of the problems identified earlier, necessitating a significant investment that Jean Francois is not convinced would be justified. Instead he would prefer to reduce the percentage of OTA bookings by closing them out to reservations during busy times, and putting higher room rates there at other times to make booking there comparatively less attractive.

Having closed during the Covid-19 pandemic, the hotel is now reopening and wants to best position itself to profit from the forecasted upswing in leisure travel. The question remains as to how best mange its distribution so as to best drive occupancy while at the same time maximising profitability. Should the hotel continue to work actively with its increasingly powerful OTA partners? Or should it try to drive more direct business by putting less attractive room rates on OTA channels and trying to replace any loss of business with online marketing efforts? In either case, in the short and medium terms, what should the hotel's priorities be?

Further reading

Abdullah, S., Van Cauwenberge, P., Vander Bauwhede, H., & O'Connor, P. (2021). The indirect distribution dilemma: assessing the financial impact of participation in Booking.com for hotels. *Tourism Review*, DOI:10.1108/TR-03-2020-0101

O'Connor, P. (2020). Digital transformation: The blurring of organizational boundaries in hotel distribution. In M. A. Gardini, M. C. Ottenbacher, & M. Schuckert (eds.), *The Routledge Companion to International Hospitality Management*. New York: Routledge. DOI:10.4324/9780429426834.ch11

O'Connor, P. (2016) Distribution channel management in hotel chains. In M. A. Gardini, M. C. Ottenbacher, & M. Schuckert (eds.), *The Routledge Handbook on Hotel Chain Management* (pp.251–253), New York: Routledge. DOI:10.4324/9781315752532.ch21

Case 14

TASTING WINE OR EXPLORING THE WINE MYSTIQUE

Aise Kim

Duration

50–60 minutes for class discussion excluding website search and/or a field trip to a winery.

Learning objectives

Upon completing the case, participants will be able to:

- understand experiential concepts which are applicable to the wine tourism context
- analyse and evaluate emerging characteristics of wine tourist markets based on a market segmentation approach
- discuss and assess the ways in which the wine tourism industry re-designs tourism services or products as nuclei for wine tourist markets with different needs or motivations
- show capacity for problem solving and analytical thinking in developing personalised tourism services for target markets.

Target audience

The case aims to help students to improve their understanding of a market segmentation approach and a critical analysis of matching target markets' needs with specific wine tourism experiences. In higher education and vocational programmes, students can apply basic marketing theories and tourism concepts to specific wine tourist markets and wine tourism product design. The market segmentation and experience design concepts are applicable to all types of tourism marketing or destination management programmes, especially for those where the focus is

DOI: 10.4324/9781003182856-14

on improving students' analytical and problem-solving skills. In particular, this integrated approach is more applicable to students who have already have background knowledge on the tourism system, tourism marketing and destination management.

Teaching methods and equipment

This case involves a review of key concepts, a case study analysis, group discussion and website searching, or field trips. Especially, the group discussion is required as it will help students to identify and compare different preferences for wine tourism experiences (e.g. tasting wines, cultural experience, or recreational activities) between group members. This will broaden up their understanding of the importance of the market segmentation approach to the tourism experience design. To complete the final task for comparing key differences and similarities in wine-related products or tourism activities between two wineries, participants are also encouraged to use their laptop to search the second case study example of a winery which offers similar or different tourism experiences for different types of wine tourist segments. Alternatively, the field trip to a winery is recommended to compare the selected case example with their own winery experience.

Teaching instructions

There are the four steps which need to be executed with a group of four to five students to complete the market-experience-product design using a case study example of the d'Arenburg winery in South Australia. In addition, students are also encouraged to search for a similar or different winery example from around the world on the internet and outline a brief description of one winery example which includes key wine tourism attractions, products and services. The four step-based discussion exercise will take up to 50–60 minutes of time for a small group discussion, and then students are encouraged to summarise and present the key findings drawn from their group discussion and compare its outcomes with other group members.

Step 1

Students are required to read key concepts described in the following sections. Part 1 (requires 5 minutes to read) describes three types of wine tourist segments who demonstrate different levels of wine knowledge and wine-related motivations. The follow-up task involves 10 minutes of group discussion, during which students are provided with the following information:

- It is important that there is at least one student selecting a different market segment outlined in Part 1 in each group (wine lovers, wine enthusiasts, wine novices), which is different from other group members' choices. In this way, each of the group members can explain and compare key characteristics of their chosen market segment to another group member.

- All group members need to explain their own preferences for the types of wine-related trip motivations and compare the differences between them and other group members' preferences.

Step 2

Part 2 introduces two major wine tourism experiences (e.g. tasting experience and cultural experience) which can be designed in a different way by wineries or wine tourism business operators. Participants need to understand key concepts relating to the physiology of taste, the definitions of wine mystique and wine terroir which play an important role in making distinctive wine tourism experiences. After reading, students are required to discuss their answers to the following questions. The group discussion could be given around 15–20 minutes.

1. Explain the role played by the senses in the appreciation of wine.
2. Identify differences in taste sensitivity and wine preferences between you and your group members.
3. Discuss how wine tourists' preferences for tasting experience differ across the three targeted wine segments (e.g. wine lovers, wine interested and wine novices).

Step 3

The case study section (Part 3) explains how the d'Arenberg winery has developed different tasting-oriented services depending on the target markets' needs. The case also illustrates how the wine mystique and wine terroir concepts are embedded into creating a unique wine tourism experience at the case study site. Instructors should give some time for students to read the case study example in detail. Students are allowed to search more information from the d'Arenberg winery website. To implement the tasks of Step 3, the market-experience-product design needs to be analysed and matched together for targeting a specific market segments' needs, motivations, or preferences. This step should last around 20–25 minutes. The following tasks are required as below.

The market-experience-product matching

Students should follow the below steps:

- *Select one wine tourist segment*: Students need to select one specific wine tourist market segment outlined in Part 1, e.g. wine lovers, wine interested and wine novices. A quick reminder of Part 1 is required to identify specific motivations of each segment and their preferences for tourism activities. For example, wine lovers are strongly motivated to focus on tasting or wine-oriented learning activities when visiting wineries, while casual wine novices are more interested in engaging in recreational activities with an emphasis on cultural entertainment or sense of place.

- *Select one tourism experience concept*: Students need to select one specific relevant academic concept relating to tourism experiences (e.g. tasting experience, cultural experience or recreational experience). It requires explaining how specific concepts (e.g. wine mystique, wine terroir) can be applicable to enhance each segment's preferences for a selected tourism experience (e.g. cultural experience).
- *Product examples*: Participants are required to select specific products/attractions/services which are available at the case study site, the d'Arenberg winery for a selected tourist experience and explain which types of specific products/attractions could satisfy a selected wine tourist segment's preferences for the selected experience (in other words, wine lovers' preferences for tasting services/products would be different from wine interested tourists' preferences).

Step 4

Step 4 is integrated with the Step 3 exercise on the market-experience-product match design. This exercise is designed to compare how the winery oriented cultural experience can be differently developed through different product/service offerings between two winery examples. Students are required to search for another particular winery around the world on the internet and identify a description of key wine attractions, wine-related products, facilities, services and other resources listed in the second winery case study site. Instructors should encourage students to analyse some similarities and differences in products/services which are offered to tourists to enhance/improve a selected tourist experience for a specific market segment's preference.

Table 14.1 can be used by students to outline key points based on their analysis of two case study examples with regards to wine tourist segments, tourism experience and products/services.

Case

Part 1: Three main types of wine tourist segments

Charters and Ali-Knight (2002) identified four segments of wine tourist markets – wine connoisseurs, wine lovers, wine interested and wine novices – depending on their level of wine knowledge and wine-related interests. Of these, three wine market segments show distinctive characteristics and these segments are well applied to many wine tourism market segmentation studies (e.g. Alebaki & Iakovidou (2011). For example, wine lovers are well educated about wine, interested in reading books on wine, have previously participated in wine tastings, and tend to

TABLE 14.1 Analysis of Case Study Examples

Task 1 – Target market segments: Select ONE tourist segment	*Task 2 – Tourism experience: Select ONE tourism experience (e.g. tasting experience or cultural experience)*	*Task 3 – Case study example 1: Products/ attractions/ services/activities which are available at the d'Arenberg winery*	*Task 4 – Example 2: Products/services/ activities from another winery selected by your group members (e.g. Jacob's Creek in Barossa Valley, South Australia)*
Wine lovers: Wine interested tourists: Wine novices:	**Option 1:** If tasting experience is selected (applying tasting preferences to each segment's needs)		
Wine lovers: Wine interested tourists: Wine novices:	**Option 2:** If cultural experience is selected (applying key concepts – the wine mystique, wine terroir, or other cultural tourism activities to the target segment's preferences)		

buy more wines during their visit to wineries. Their motivations for visiting wineries also include learning about wine-related history, tasting a higher quality of wine (e.g. premium wine) or specific brands of wine, and food-wine pairing classes. Wine lovers are less interested in other tourist activities, such as recreational tours or entertainment-oriented activities. The wine interested tourists are more likely to have attended a tasting at a winery before, but they are not educated about wine or do not have high levels of wine knowledge or tasting skills. This segment tends to be interested in learning about how to taste different kinds of wine, and engages in wine-related tour activities, such as learning about wine-making process, wine storing and ageing, and wine-relating farming process associated with soils and climate. However, the wine interested tourists are less interested in food and wine links. The final segment, the wine novice, is considered as curious tourists who are interested in tourism-related activities (e.g. cultural or recreational activities) at wineries rather than tasting or learning more about wine itself. They tend to have little knowledge of wine or be less interested in wine tasting, but this segment is more interested in exploring a sense of place around a winery environment than other two groups.

Part 2: Wine tourism experience

Tasting experience

Tasting experience has been identified by many researchers as being a primary motivation for wine tourist markets. Taste is a means by which we determine if something is edible and palatable (Brillat-Savarin, 1970). According to the physiology of taste, people have specific sensitivities of taste buds which lead to having different personal tasting preferences for certain types of food or wine (Brillat-Savarin, 1970). When tasting wine, consumers use different senses, such as flavour, smell, or colour, to evaluate the quality of tastes or suitability of tastes to their preferences. For example, wine tourists' preferences for tasting specific types of wine could be influenced by various aspects of wine attributes with added flavours of blackberry or spices. The texture of wine is also important in determining the tasting process which can include the weight, tannin and alcohol of the wine. The temperature of the wine also can impact on the taste of the wine. The right temperature of chilling white wine could be a critical factor for satisfying consumers' taste buds. Otherwise, further acidic taste could be intensified due to chilling it too much (Laloganes, 2010). Previous research goes further to confirm that younger women tend to prefer sweet or white wine, while older women tend to prefer red wine. By contrast, both younger and older men tend to drink stronger red wine such as Merlot, rather than sweet or white wine (Bruwer, Saliba, & Miller, 2011). This implies that tourists' tasting preferences can be distinguished depending on their level of wine-related knowledge and their tasting skill levels as well as gender or age differences. Thus, wine tasting experience requires a careful and deliberate matching approach to the development of wine selection and food and wine pairing for meeting specific needs of wine market segments.

Cultural experience: Wine mystique and terroir

Wine tourist markets are also motivated to explore cultural aspects of wineries and wine-related history when visiting wineries. Each winery brings out the opportunity for many authentic encounters with wine-related history and winery's local environment which could be appealing as tourism attractions. Cultural tourism experience at wineries can be created in a different way by identifying the wine mystique or the regional terroir as part of the winescape attributes. Wine mystique refers to the mystical power that surrounds wine. Unique wine-related history or heritage could intrigue, making wine more desirable or mysterious, along with the pleasure of the wine itself. The terroir of wineries is associated with a set of aesthetic, social, cultural and regional attributes (e.g. the landscape, regional climate, local people's lifestyle) that appear to be attractive to tourists. The terroir (the French word for 'soil') refers to the local environment and the landscape that adds to the mystique of wine tourism experience. Utilising the wine mystique, wineries can develop a unique tour of the cellar door to provide tourists with a learning opportunity to discover how the wine is being made (showcasing all the stages of making the wine to it ending up in the bottle), while allowing tourists to taste different kinds of wine made over various years and using different techniques. Along with the natural features of the local wineries, wine tourist markets can expand their unique authentic cultural experience to exploring how grapes grow in what type of soil and what type of climate is required for wine farming. Apart from the winery's winescape attributes, the rural settings associated with heritage features around local winery destinations are attractive to tourists who are interested in participating in the cultural activities of a rural destination (e.g. people in the local community, historical buildings, local people's lifestyles, or arts).

Part 3: A description of a winery case study

The d'Arenberg cube building has become an iconic wine tourism attraction which is situated within the d'Arenberg vineyards in McLaren Vale winery region in South Australia. This cube-shaped building concept was originally inspired by Chester Osborn, a Chief Winemaker of d'Arenberg vineyards, representing the complexities and puzzles of winemaking. He is of the fourth generation of the Osborn family, who established the vineyards in 1912. Since its opening in 2017, this building has become known for its distinctive geometric design, resembling a Rubik's Cube. This five story building features multi-use facilities such as the museum, tasting room for wine tourists, cellar doors and restaurant. Visitors are encouraged to explore a winery sensory room, a virtual fermenter, a 360-degree video room, other sensory experiences, and the Alternate Realities Museum with a focus of modern art installations.

The d'Arenburg winery offers various tasting experiences, ranging from tasting entry or artisan wines to creating personalised blending wine experience. The tasting room is located at an upscale restaurant on the top level of the d'Arenburg Cube building, highlighting a 360 degree panoramic winery view over McLaren Vale. With a cost of US$15, it allows visitors to taste different kinds of wine and visit the Alternate Realities Museum, located on the ground floor of the building. Through tasting experience, visitors are able to distinguish the style and texture of wine. The d'Arenburg's white wine collections guide visitors to explore expressive wines with lifted aromas and delicate bouquets, or full flavoured wine with fresh acidity and a long lingering finish. For tasting red wines, visitors would be able to distinguish different textures of red wines which present various styles ranging from fragrant, fruit flavoured artisan wines with excellent texture or a hint of oak flavour, to the iconic wine with a great capacity for longevity, and to the super premium wine with elegance and finesse. Recently, as a new tasting experience, the d'Arenburg winery also offers a hands-on experience which stimulates tourists' sense of adventure, blending different flavours of wine, using three barrel samples of single vineyard wine, and then taking home a personalised bottle of Shiraz wine. This creative tasting experience provides tourists with luxury and culinary elements through a mixture of masterclasses and a food and wine pairings course, called 'à la carte or degustation dining' experience at d'Arry's Verandah Restaurant. This upscale restaurant is well-known for showcasing the best views overlooking the rolling hills of McLaren Vale. The restaurant's menu represents the terroir of McLaren Vale, showcasing local and seasonal produces, which include Pete Reschke's signature lobster dish, cheese and coffees. For the last, but best, highlight of the winery experience at this cube-shaped building, visitors have a great opportunity to explore a contemporary art museum with various art exhibitions, or participate in helicopter rides with outstanding views around McLaren Vale.

The d'Arenberg winery provides visitors with a great opportunity to discover a unique winery identity which is associated with traditional wine-making techniques, such as open fermentation, foot treading and basket pressed wine before fermentation. Much of the traditional basket pressed wine-making process has not been practiced due to the labour-intensive process. Yet, d'Arenberg is one of the only wineries in Australia to basket press all wines, making this traditional technique worthwhile, resulting in the best quality of wine. For instance, this basket pressed white wine making process helps to ensure no colour or tannins are extracted from the skins of grapes. For red-wine, the unique traditional technique relies on a slower wine-making process allowing appropriate aging, while also reducing oak influence. Other wine-making processes can be also discovered by visitors who learn more about how and when the grape-pick time is determined by wine makers, how small batches are basket pressed and transferred to a mixture of new and used French oak barriques to complete primary and secondary fermentation, and how foot treading is undertaken during fermentation. An extensive barrel tasting process is undertaken by the wine making team to determine the entry level of wine or rare wines. With a growing concern of sustainability, d'Arenberg wineries also produces NASAA certified wines based on organic and biodynamic processes, with a strong commitment to achieving the goals of environmental sustainability for maintaining the soil quality and the winery environment.

(Note: The information presented in this case were extracted and revised from the d'Arenberg website. For further details, visit www.darenberg.com.au/darenberg-cube).

References

Alebaki, M., & Iakovidou, O. (2011). Market segmentation in wine tourism: A comparison of approaches. *Tourismos: An International Multidisciplinary Journal of Tourism*, 6(1), 123–140.

Brillat-Savarin, J-A. (1970). *The Physiology of Taste*. London: Penguin Books.

Bruwer, J. Saliba, A., & Miller, A. (2011). Consumer behaviour and sensory preference differences: Implications for wine product marketing. *Journal of Consumer Marketing*, 28(1), 5–18.

Charters, S., & Ali-Knight, J. (2002). Who is the wine tourist? *Tourism Management*, 23, 311–319.

Laloganes, J. P. (2010). *The Essentials of Wine, with Food Pairing Techniques*. Upper Saddle River, NJ: Prentice Hall.

Further reading

Charters, S. (2006). *Wine and Society: The Social and Cultural Context of a Drink*. Oxford: Butterworth-Heinemann.

d'Arenberg (2021). The d'Arenberg Cube. www.darenberg.com.au/darenberg-cube

Hall, C. M., Sharples, L., Cambourne, B., & Macionis, N. (2000). *Wine Tourism Around the World: Development, Management and Markets*. Oxford: Butterworth-Heinemann.

Case 15

DESIGNING GASTRONOMIC IDENTITY-BASED FOOD TOURS

Aise Kim

Duration

Approximately between 50–60 minutes.

Learning objectives

Upon completing the case, participants will be able to:

- understand the gastronomic identity concept which is applicable to the food tourism context
- identify and analyse various food tourism experiences and relevant food tour examples which are connected to food tourists' emerging demand trends
- discuss and critique the ways in which the food tourism industry develops distinctive/unique food tourism experiences using various products and tourism activities, for tourist markets with different needs or motivations
- enjoy enhanced creative engagement activities through an analysis of new trends and emerging demand patterns.

Target audience

This case targets students who have the background knowledge on tourist motivations, tourism experience, tourism marketing and new product development. In higher education and vocational programmes, students can apply basic marketing theories and tourism motivation concepts to analysing specific food tourism experiences and new food tourism product design. In particular, the case aims to help students to identify tourist motivations for food tourism experiences and analyse how such food-oriented motivations are embedded into new tourism product

DOI: 10.4324/9781003182856-15

development. Based on the analysis of new trends or challenging issues in the development of food tourism products, students are also expected to demonstrate their creativity-based problem solving skills which reflect current demand trends for unique food tourism experiences.

Teaching methods and equipment

To complete the required tasks, students need to review the case study, analyse key motivational concepts, apply key experiential concepts to the case study examples and create new food tours. Through the group discussion and website searching (or along with the field trip to the local markets), students are required to demonstrate their creative ideas for recommending new tour activities which could help improve the current food tour operators' tourism business development while overcoming the challenging issues or weaknesses of their tour business (e.g. similar product offerings or a narrow-focused niche market approach). This exercise will broaden students' understanding of the importance of the gastronomic identity approach to the tourism experience design. Furthermore, with the focus of food tourist markets, this study will strengthen students' knowledge application and problem-solving skills, developed in Case 15 on wine tourist markets.

Teaching instructions

Four steps need to be executed with a group of four to five students to analyse the strengths and weaknesses of the food tour design using a case study example of Food Tours Australia in Adelaide, South Australia. For an alternative option, instructors can encourage students to search for a similar or different food tour example from around the world on the internet and outline a brief description of this additional example, which should include unique or innovative food tour ideas appealing to food tourist markets' emerging demand trends. The three step-based discussion exercise will take up to 30–40 minutes of time for a small group discussion. As for the last step, students should demonstrate their creative ideas in a competition for the best unique food tour experience, using a drawing poster or summary notes. All students can discuss and make a judgement on which group's creative ideas are the best solution for improving the case study example's food tourism business. This creative idea presentation session will take 10–15 minutes.

Step 1

Part 1 introduces the gastronomic identity concept and outlines various types of food tourism experiences (e.g. tasting experience, cultural experience, prestige, sense of place) which can be designed in different ways by food tourism business operators. Participants need to understand how the gastronomic identity could be redesigned, using local food/ethnic food products in order to make distinctive cultural experiences for food tourists. After reading Part 1, students are required to discuss their answers to the following questions. The group discussion could be given around 10–15 minutes.

1. Analyse different aspects of ethnic food or local food which determines unique gastronomic identity and values.
2. Compare different examples of ethnic food tour activities which enhance distinctive gastronomic experiences.

Step 2

As outlined in Part 1, Step 2 aims to identify various types of food tourism experiences which are connected to emerging food tourists' motivations. This task involves 5 minutes of reading one section of the tourism experience described in the following selected journal article:

> Kim, Y.G., Eves, A., Scarles, C. (2009). Building a model of local food consumption on trips and holidays: A grounded theory approach, *International Journal of Hospitality Management,* 28, 423–431.

Students are expected to read only one tourism experience section (subheadings from '3.2. Escape from routine', '3.3. Health concern', '3.4. Learning knowledge', '3.5. Authentic experience', etc. to '3.9. Physical Environment' and '3.10. Physiological factors') listed in pages 426 to 428. Instructors need to allocate each of the group members with a different section of reading – for example, the first student from the group is allocated with '3.7. Prestige' but the second student with '3.8. Sensory appeal'. Each participant is assigned one particular tourism experience (e.g. '3.5. authentic experience') from the reading listed above which is different from other group members' choice. In this way, each of the group members can explain and compare specific food tourism experiences and relevant examples of tour products/activities to another group member. After reading, students should spend 10 minutes discussing the following question:

3. Explain a selected food tourism experience, with relevant food tourism examples, to your group members.

Step 3

The case study section (see Part 2) explains various food tour activities and products developed by the Food Tours Australia company. The case also illustrates how various elements of the gastronomic identity are embedded into the food tourism experiences at the case study site. Instructors should give some time for students to read the case study example in detail. Students are allowed to search more information from the Food Tours Australia website. This step should last around 20–25 minutes. The following tasks are required:

- *Select one tourism experience concept*: Students need to select one specific relevant academic concept relating to tourism experiences (e.g. tasting experience, cultural experience, learning experience, health-oriented experience, sustainability-based experience, or sense of place experience). It requires explaining how specific food-related values/activities (e.g. local food identity, ethnic food) can be applicable to meet tourists' specific motivations for a particular tourism experience (e.g. cultural experience).
- *Product examples*: Participants are required to categorise and analyse specific products/attractions/services provided by the case study example, 'Food Tours Australia'. Students need to assess which types of specific products/attractions could satisfy food tourists' selected tourism experiences (in other words, food lovers' preferences for tasting experience or learning experience for practicing cooking skills through blind tasting).

Students should then answer the following questions:

4. Analyse different aspects of gastronomic identity and values associated with ethnic foods which are available at the case study site, Adelaide Central Market.
5. Identify some strengths and weaknesses of the food tour activities provided by the case study example (Food Tours Australia), in terms of emerging food tourism trends, new food tourists' preferences, similar food tour activities or missing food tourism experience examples.

Step 4: Creative idea competition

Step 4 is a creativity-based task building on the Step 3 exercise on the market-experience-product match design which is similar to the winery example in Case 15. Instructors should encourage students to analyse other food tour examples or visit the local markets (as part of the field trip) which inspire students' creative ideas for promoting unique food tourism experiences for relevant tourist markets. This exercise is designed to compare how the food tourism experience can be developed differently through innovative or unique product/service offerings. Set students the following question:

6. Recommend a new product or tour activity which enhances a selected tourism experience using gastronomic identity, new product ideas and servicescape concepts for the unique tourism experience.

Students are required to recommend new products/services which could be offered to a new tourist market to enhance/improve the unique food tourism experience, while overcoming the weaknesses of the current food tour business, Food

TABLE 15.1 Analysis of Case study Examples

Task 1 – Tourism experience: Select ONE tourism experience (e.g. prestige, learning experience, cultural experience or sense of place)	*Task 2 – Case study example 1: Analyse products/attractions/services/activities which are provided by 'Food Tours Australia'*	*Task 3 – Creative ideas: Recommend new products/services/activities for enhancing your target market's desired experiences*
Gastronomic cultural experience Sense of place experience Learning experience Other type of food tourism experience		

Tours Australia. Table 15.1 can be used for students to outline key points based on their analysis of the case study example with regards to food tourism experience, products/services from the case study example and creative ideas.

Case

Part 1: Gastronomic identity and food tourism experience

There has been increasing interest in redeveloping various local foods and adding new values to the local food identity as gastronomic identity can act as an important differentiator between destinations. This gastronomic identity-based product design helps tourism business providers to develop a distinctive or unique tourism experience, while avoiding similar product offerings. Gastronomic identity is defined as 'the influences of the environment (geography and climate) and culture (history and ethnic influences) on prevailing taste components, textures and flavours in food and drink' (Harrington, 2006, p.130).

The model in Figure 15.1 explains that the gastronomic identity of a specific country's food is influenced by a broader aspect of the local environmental factors such as culture, geography or locally produced agricultural products. That is, the ingredients of traditional cuisines can differ between regions, and have an influence on the flavours, recipes, cooking styles, mixture of ingredients and dining etiquette that form a part of the gastronomic identity (Harrington, 2006). Recently, gastronomic identity is also influenced by the ethnic diversity of the country, new technology or changes in servicescape offerings. For example, the fusion of Australian food represents a unique and contemporary Australian culinary identity, as Australia's fusion food has been developed under the influence of migrants' ethnic food, which is combined with Australian's mainstream food culture. Such a fusion food diversity shows a mixture of indigenous food ingredients with European style of cuisines, or combining Australian local products with Asian cooking styles and flavours. The availability of different ethnic cuisines or fusion food in Australia

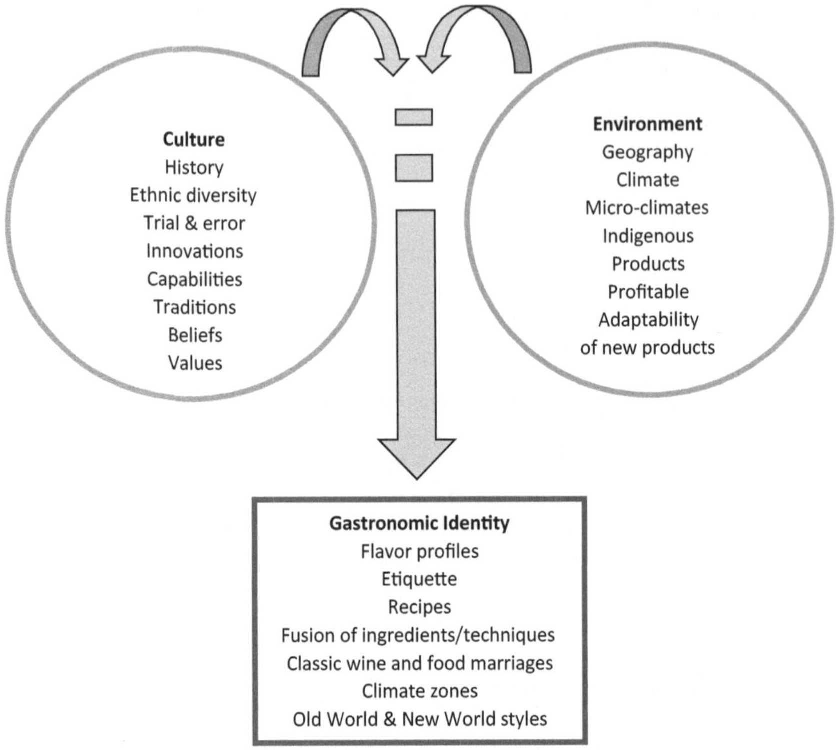

FIGURE 15.1 Gastronomic Identity Model

Source: Harrington, 2006, p.130

could be considered as a unique cultural experience as it can play an important role in introducing tourists to new flavours, different traditions, culinary etiquette or ways of eating which differentiate one region from another. Thus, tasting local food or fusion food with an emphasis on the unique gastronomic identity could serve as both a cultural activity and entertainment, as it could enable tourists to experience a distinctive culture of a particular society (Fields, 2002).

Beyond the cultural experience of food tourism, recently, food tourism researchers have identified various types of food tourism experiences which are also entwined with the local destination's culture and environment or local people's values/beliefs. The diversity of food tourism experiences reflects new demand trends which are increasingly being sought by current food tourists. Kim, Eves, and Scarles (2009) categorised key food tourism experiences into three contexts: 'motivational factors' (i.e. exciting experience, escape from routine, health concern, learning knowledge, authentic experience, togetherness, prestige, sensory appeal, physical environment); 'demographic factors' (i.e. gender, age, education); and 'physiological factors' (i.e. food neophilia and food neophobia). For example, prestige, health concerns or interpersonal motivators have become one of new demand trends as current food tourists want to be seen in the 'right' or 'trendy' upscale restaurant and indulge in a high quality food experience because it reflects their status or lifestyle. Another example is food lovers' active engagement in cooking schools or touring around local markets which are considered part of authentic learning experience for food tourist markets.

Part 2: Adelaide Central Market Tour, Adelaide, South Australia

'Adelaide Central Market Classic Discovery Tour' is offered by Food Tours Australia, an award-winning food tourism company in South Australia. Food Tours Australia unravels the secrets and stories of Adelaide Central Market with its long history of 150 years since opening in 1869. This central market place has become an iconic culinary attraction for appealing to local food lovers. Started by a food tour operator, Mark Gibson – ex-chef, market stallholder and food expert – this market-based tour highlights a picturescape and busy atmosphere of the market place encompassing over 80 small artisan food stalls or shops which specialise in fresh, locally grown and produced foods.

Adelaide Centre Market-oriented tours focus on a premium tasting experience through exploring various types of local products, such as artisan cheese, honey, bush tucker, quirky Australian native foods, seasonal vegetables and fruits, seafood, dairy products and ethnic food, while meeting artisan food producers, suppliers and stallholders and learning more about their life stories relating to trading at the market for generations.

The tastescape of the central market introduces tasting fresh local flavour profiles and ethnic food diversity, while dining at the café, or enjoying a cooking demonstration on a special occasion. Along with locally produced cheese varieties and seafood, unique local products and natural food are also introduced to visitors. For example, the 'something wild' shop showcases unique Australian native foods (including gin combined with green tree ants) which are picked from the natural environment and eaten by Indigenous Australians because of the nutritional value and health benefits of these superfoods. Another example is the 'Kangaroo Island' stall, which showcases a unique combination of local products by pairing gins and honey, pig pastes or natural plant-based hand creams.

As for another example, a special cheese masterclass tour targets food lovers or food interested tourists who would be keen to taste various types of cheese from across the world ranging from locally made artisan cheeses to Cheddar varieties, blue cheese, French Bries, Spanish Manchego, or Italian Gorgonzola. Other learning activities are also provided to tourists as they can learn more about preparation, presentation, manufacturing, handing, and storage of cheese from a cheese expert. The guide gives tourists a chance to compare different tastes between mass produced and artisan produced cheese products.

The Adelaide Central Market-based tours are also extended by diversifying specific target audiences which include educational food tours for school groups, corporate tours, exclusive private tours, unique Australian local food or ethnic food-oriented tours for both domestic and international tourists. These educational food tours introduce ethnic food diversity, which represents the multicultural society of Adelaide as migrants' food culture has become part of the contemporary Australian gastronomic identity. The educational food tours have been developed by Food Tours Australia with a focus of four main themes – Italian food, Chinese food, African food and Indonesian food (e.g. Nasi goreng). Each ethnic food-focused education tour offers visitors not only the chance to learn about iconic ethnic cuisines and food culture but also language, music, dance, personal stories from tour guides or migration history to South Australia.

Additional entertainment activities are added as part of fun-oriented cultural experiences including wearing Venetian masks, replica Chinese school uniforms, African headwear or original headgear from Indonesia. For example, Italian speaking guides offer students a great opportunity to practice their Italian when starting the tour. While wearing replica Venetian masks during the tour, students can taste Italian delights such as olives, ciabatta, salamis, pasta dishes, desserts etc. Such a tasting experience helps students to develop a better understanding of Italian cuisine available in Adelaide, whilst learning more about the local people's lifestyle and Italian migration history to South Australia. For the African food tour experience, Year 4 Geography students can expand their appreciation of a unique African cultural experience through African language, food, music and dance, while tasting African foods such as merguez sausages, ingera and beef jerky across the stalls.

Apart from the ethnic food-oriented tours, Food Tours Australia has also developed a different theme of educational tours for targeting specific food sustainability issues. Students are educated by a sustainability expert about organic farming, carbon footprints and food miles-related issues, while highlighting the important value of artisan produced items or organic food available at the markets, compared to mass produced products. Furthermore, a short part of the field trip also focuses on learning about food-related economics. This part of the tour is based on practical learning activities by analysing costs of ingredients, and practicing the most cost-effective way of shopping considering the benefits of health, nutrition and sustainability aspects of their food choice, and then cooking the food items bought from the market and sharing the meal together at the Market Kitchen, which has recently been built at the Central Market. Considering an extended market segment, another educational food tour is targeted for corporate groups or younger groups. With a slogan of 'Think you know food?', this tour challenges food interested tourists' culinary skills and their agility in high intensity sensory skirmish, creativity, food preparation or teamwork. Exciting hands-on activities such as blind tasting, or treasure hunts around the market in 90 minutes are a unique highlight for this tour.

Recently, 'Food Tours Australia' also started offering a sense of place experience around the Adelaide city environment by linking up with city icons, landscapes and history, along with a culinary delights experience at the Adelaide Central Market. The recreational activities with bike tours or eco-cap tours are associated with the Central Market. It shows a diversification approach to mixing different activities around the city environment. This tour combines the best of Adelaide's foodie culture, east and west end bar and restaurant districts, the riverbank, heritage sights, sports and arts districts as well as the northern parklands to get around town, while hopping on or off Segways or bikes. Another tour introduces Adelaide Showground Farmers' Market, the largest farmers' market in South Australia. Visitors have a chance to explore fresh and seasonal products and agricultural diversity representing each region of the state, while having a sense of local people's lifestyle through vibrant interactions between small family farmers/producers, and local people's shopping experience every Sunday.

Note: The information in this case study was extracted and revised from the Food Tours Australia website. For further details, visit www.ausfoodtours.com/list-of-tours.

References

Fields, K. (2002). Demand for the gastronomic tourism product: Motivational factors. In A.-M Hjalager & G. Richards (Eds.), *Tourism and Gastronomy* (pp.36–50). London: Routledge.

Harrington, R. J. (2006). Defining gastronomic identity: The impact of environment and culture on prevailing components, texture and flavours in wine and food. *Journal of Culinary Science and Technology*, 4(2/3), 129–152.

Kim, Y. G., Eves, A., & Scarles, C. (2009). Building a model of local food consumption on trips and holidays: A grounded theory approach. *International Journal of Hospitality Management*, 28, 423–431.

Further reading

Boniface, P. (2003). *Tasting Tourism: Travelling for Food and Drink*. Hampshire, UK: Ashgate.

Quan, S., & Wang, N. (2004). Towards a structural model of the tourist experience: An illustration from food experiences in tourism. *Tourism Management*, 25, 297–305.

Richards, G. (2002). Gastronomy: An essential ingredient in tourism production and consumption. In A.-M. Hjalager & G. Richards (Eds.), *Tourism and Gastronomy* (pp.3–20). London: Routledge.

Case 16

GLOBAL HOTEL REVENUE MANAGEMENT CHALLENGES

A nightmare for revenue managers

Tevfik Demirçiftci

Duration

Between 55–65 minutes.

Learning objectives

Upon completing the case, participants will be able to:

- explain the concept of RM and the history of its development
- define total hotel RM
- discuss the challenges in total hotel RM.

Target audience

The purpose of this case study is to explain RM concepts that were taken from a real-world situation. This case study can be used in hospitality and tourism vocational programmes and higher education. It is suitable for both individual and group work. The ideal class size can is between four and ten. This case study will help students understand what RM is, learn the duties of revenue managers, identify the global challenges facing revenue managers, and it will recommend tips for future revenue managers. The author conducted a Zoom interview with Mrs Deniz Dorbek Kocak – an RM expert – on 4th April 2021. The duration of this interview was 65 minutes. In this case study, Mrs Kocak's career is explained in detail to help students consider a career in hotel RM.

Teaching methods and equipment

This case study will be used for RM classes. Before the class, students are asked to read the case study. A cc omputers and projector will be required, and several

DOI: 10.4324/9781003182856-16

questions can be asked about RM on game-based learning platforms, such as Kahoot. Before the case study, a traditional lecture needs to be delivered by the instructor. This case can be used as a class discussion exercise supporting the 'Introduction to RM' class. The class can be divided into small groups (two students per group), and each student group can be assigned one question to answer. The groups can then share their answers with the class. At the end of the class, several multiple-choice questions can be asked through Kahoot. Instructors might invite guest speakers to the class.

Teaching instructions

Step 1

The following video clip about hotel RM should be shown to the class:

> www.youtube.com/watch?v=b2zg81CSZ64

It might be a suitable start to discussing RM. Instructors can explain several real-life examples related to RM. Stay-three-pay-two room promotions, room package promotions and two-for-one restaurant promotions can be listed as examples in the beginning. Students should be encouraged to provide some examples related to RM (5 to 10 minutes).

Step 2

Start the class with brief information about hotel RM. First, a brief history of RM can be discussed in a traditional lecture format. Furthermore, several criteria for practical usage of RM can be mentioned. These criteria are fixed capacity environment, perishable products, varied but predictable demand, high fixed costs and low variable costs. Students might have difficulty with the criteria 'various but predictable demand'. To overcome this challenge, the example below can be provided. A warm winter resort in Miami might have different seasonal demand fluctuation from that of a business hotel in New York. Thus, every hotel has a different demand pattern (10 minutes).

Step 3

Lecture about RM for the lodging industry. Students will learn how to calculate key performance indicators for hotel RM (Table 16.1). This information will be based on hotel RM applications such as competition check, SWOT analysis, forecasting, distribution channel management and performance analysis. Several performance indicators, such as revenue per available room (RevPAR), occupancy percentage (OCC%) and average daily rate (ADR), should be discussed. Microsoft Excel should be used to calculate key performance indicators. Students who are not familiar with Excel might have problems while calculating key

TABLE 16.1 Revenue Management Key Performance Indicators

Occupancy rate = number of occupied rooms/total number of available rooms
Average daily rate (ADR) = rooms revenue/number of rooms sold
Revenue per available rooms (RevPAR) = rooms revenue/rooms available or average daily rate x occupancy rate
Gross operating profit available per room (GOPPAR) = gross operating profit/available rooms
Total revenue per available Room (TRevPAR) = total revenue/number of available rooms

Source. 'Revenue Management KPIs' by Revfine (www.revfine.com/category/revenuemanagement-kpi/)

performance indicators; instructors need to help such students during the class. At the end of this phase, students will have learned how to calculate key performance indicators.

Step 4

Ask each student or group to summarise the case study by answering the questions below. Divide the class into small groups (two students per group). Give each student group (or individual student) one question to answer. Ask the students to share their answers with other groups (30 minutes).

1. What is the case study about?
   It is about the challenges in total hotel RM and global hotel RM.
2. What is Deniz Dorbek Kocak's background?
   Her RM-related work experience, such as working with Hilton, Wyndham, and Movenpick groups.
3. What is total hotel RM?
   It is an approach that integrates all revenue streams of a hotel property to optimise revenue.
4. What are the challenges in global RM?
   - Technological challenges
   - Organisational challenges
   - Financial challenges
5. What are Deniz Dorbek Kocak's final recommendations to students?
   - Hospitality is a people business.
   - Students need to follow the trends in the hospitality industry and economic, social and technological environments.
   - Students need to be open to travelling.

At the end of this phase, students will have understood the challenges in global hotel RM.

Step 5

Kahoot questions (10 minutes).

1. In which hotel group did Mrs Deniz Dorbek Kocak not work?
   a. **Marriott (correct answer)**
   b. Hilton
   c. Movenpick
   d. Wyndham
2. What are the challenges in global revenue management (RM)?
   a. Technological challenges
   b. Organisational challenges
   c. Financial challenges
   d. **All of the above (correct answer)**
3. What is the formula for revenue per available room (RevPAR)?
   a. **Rooms revenue/rooms available (correct answer)**
   b. Rooms revenue/number of rooms sold
   c. Gross operating profit/available rooms
   d. Total revenue/number of available rooms
4. What is the formula for gross operating profit per available room (GOPPAR)?
   a. Rooms revenue/rooms available
   b. Rooms revenue/number of rooms sold
   c. **Gross operating profit/available rooms (correct answer)**
   d. Total revenue/number of available rooms
5. What is the formula for occupancy percentage (OCC%)?
   a. **Number of occupied rooms/total number of available rooms (correct answer)**
   b. Rooms revenue/number of rooms sold
   c. Total revenue/number of available rooms
   d. Rooms revenue/rooms available or average daily rate x occupancy rate
6. What is the formula for the average daily rate (ADR)?
   a. **Rooms revenue/number of rooms sold (correct answer)**
   b. Total revenue/number of available rooms
   c. Gross operating profit/available rooms
   d. Number of occupied rooms/total number of available rooms

Case

Background of Deniz Dorbek Kocak

After completing her bachelor's degree in hospitality management in Turkey, Mrs Kocak first started to work in the Hilton Hotel group as a front-office agent in Istanbul. She worked for this company in several positions in Izmir and Kayseri,

Turkey; Edinburgh, Scotland; and London, UK. Her first managerial experience was working as a reservation and yield manager in Hilton Izmir. As a result of her outstanding performance, she was promoted to business development manager in Hilton Kayseri. After spending eight years at Hilton, she moved to Thailand in 2008 and worked as a director of RM at Movenpick Hotel Resort in Thailand for two years.

After working in Thailand, she worked as an associate director of RM in Atlantis, the Palm Jumeirah in Dubai. In the Middle East, she also worked for Kempinski Hotel in Jordan between 2011 and 2012. After spending two years in the Middle East, she moved back to Turkey to work as the area director of RM for Turkey and East Europe for the Fairmont group between 2012 and 2015. After her tenure with Fairmont, she served as the vice president of global RM operations for Wyndham Hotels and Resorts for four years. She is now working as an RM consultant for major international brands globally.

Total hotel RM and challenges

Zheng and Forgacs (2017) defined 'total hotel RM' as 'an approach that integrates all revenue streams of a hotel property to optimize revenue' (p.238). Total hotel RM allows revenue managers to create revenue opportunities from food and beverage, function space, spa, retail and golf, and other revenue streams. According to Mrs Kocak, there are several challenges in the application of total hotel RM. These challenges might be summarised as technological, organisational and financial.

Technological challenges

Mrs Kocak stated that RM in the hotel industry is still new. First, the airline industry adopted RM, and then hotels and restaurants started to adopt RM. According to Mrs Kocak, one of the most critical components of RM was technology software. She mentioned that 70% of revenues have come from the rooms department. Hence, terrific RM software such as Duetto and Rainmaker have helped revenue managers maximise their room revenue in full-service hotels. She also discussed using a basic Excel file with macros to perform RM in economy-segment small hotels. However, there is no specific RM software focusing on restaurants or spas. This lack of technology was a big challenge for the hospitality industry. Thus, it was complicated for revenue managers to collect data. To overcome this problem, she recommended using data such as average guest check-in, meal duration and so forth in the point-of-sale (POS) systems of restaurants. Finally, she advised that the customer relationship management (CRM) system and RM systems needed to directly interface with each other to implement RM effectively.

Another critical concern that Mrs Kocak highlighted was that the hospitality industry was not data driven. She asserted that there was a gap between digital marketing and RM. She said that a large amount of information could be found through one's website. This information might be listed as website visitor statistics, popular check-in dates, conversion rates, confirmed bookings and average revenue per reservation. It was recommended that this information be sorted by zip code to deduce where the reservations were coming from. She also stated that artificial intelligence would be critical for the hospitality industry and noted that competition was intense within the industry, which Airbnb has made more challenging. She advised that hospitality companies needed to manage data effectively to have a competitive advantage.

Organisational challenges

The first recommendation regarding organisational barriers was that commercial departments (finance, sales and marketing) and operational departments (front office, food and beverage, and spa) needed to work together to maximise revenue. Hence, RM culture was important. As a revenue manager, one needs to train one's executive chef and restaurant manager in RM fundamentals, such as revenue per available seat hour (RevPASH), revenue centre contribution percentage and revenue per labour hour. She stressed the importance of menu engineering. Selling high-demand menu items might contribute to the bottom line effectively. Both setting menu prices and managing costs are essential for revenue managers. Finally, she noted that an upselling programme for food and beverages and front-office departments might boost revenue.

Another reported problem was that revenue managers focused only on room profitability, as hotel companies did not allocate sufficient resources. In some hotels, a revenue manager was responsible for two or three hotels. Thus, a revenue manager could focus only on room profitability. She stated that total hotel RM could be applied effectively if two or three revenue analysts assisted a revenue manager. She also added that the revenue manager was not the only person responsible for maximising revenues; all employees were responsible for maximising revenues. Therefore, she stressed the importance of RM training for executive-level and line employees in the hotel.

Financial challenges

Mrs Kocak also made some negative comments about the key performance indicators of RM practices. She stated that it was not rational to evaluate performance based on ADR, OCC%, and RevPAR only. She said that new metrics such as GOPPAR and total revenue per available room (TRevPAR; see Table 17.1) needed to be used. She noted the weakness of the profit and loss statement (income statement), and she said that it was impossible to obtain a market segment's profitability in the profit and loss statement.

Final recommendation to students

According to Mrs Kocak, the hospitality business is a people business. She noted that working in the hospitality industry would be nearly impossible if a person is not suited to working with people. She added that students must follow the trends in the hospitality industry and economic, social and technological environments. She also suggested that the hospitality industry is suitable for those who are fond of travelling. According to her, learning a foreign language is an advantage. Finally, students need to think outside the box to be creative problem solvers.

Reference

Zheng, C. and Forgacs, G. (2017). The emerging trend of hotel total revenue management. *Journal of Revenue and Pricing Management*, 16, 238–245.

Further reading

Cetin, G., Demirciftci, T., & Bilgihan, A. (2016). Meeting revenue management challenges: Knowledge, skills, and abilities. *International Journal of Hospitality Management*, 57, 132–142. https://doi.org/10.1016/j.ijhm.2016.06.008

Hayes, D. K., & Miller, A. A. (2011). *Revenue Management for the Hospitality Industry*. Hoboken, NJ: John Wiley & Sons.

Ivanov, S. (2014). *Hotel Revenue Management: From Theory to Practice*. Varna: Zangador.

Case 17

A YACHT CHARTER HOLIDAY

Serim Paker

Duration

Because the original case was intended as a term project, the duration is variable based on the students, the desired objective and the expected level of detail. Five weeks with the expectation of a high level of attention into all elements, including boats' nautical features and a realistic sail plan ,works well for post-graduate Marine Tourism students; ten days of more abstract study is appropriate for Maritime Business and Administration students. The tutor should alter the expectations as well as the duration accordingly.

Learning objectives

Upon completing the case, participants will be able to:

- understand digital marketing platforms for charter yachts
- develop a customers' point of view on the yacht chartering market
- comprehend the amenities and properties of the yacht, and be exposed to maritime English
- enhance their abilities on booking yachts, flights and other necessary bookings
- compare prices of services offered.

Target audience

This case study aims to put students in the shoes of yacht charter customers. Given the flexibility of the intended level of depth, the study can be tailored to a variety of courses in digital marketing, tourism management and marketing, maritime

DOI: 10.4324/9781003182856-17

business and administration, marine tourism and any educational environment that sees the case as relevant.

Teaching methods and equipment

The study relies heavily on class discussions and online browsing. A classroom with a data projector and an internet connection would suffice. To complete their cases, the students would need their own computers or access to a computer at a library.

Teaching instructions

The following sections outline the steps involved in the case. As mentioned before the instructor should feel free to alter the steps according to expected outcomes. It is not recommended to provide students with extensive information or to issue frequent warnings. Allowing them to make mistakes will make the study more durable and efficient. Instead of warning about the seasons and weather, let some of them slip and blunder about dates. For example, if a student plans to charter a yacht from Seychelles in January, you may show the class what the Indian Ocean islands and surrounding waters look like during monsoon season. Compared with warning people not to rent a yacht anywhere during storm seasons, this will lead to a better learning experience which will be stored in long-term memory.

Step 4 is a lengthy step, and adequate time should be provided to the students if this step matches the intended aim of the study/class. This step may be abstract if the class is not part of a maritime-focused course.

Case

Introduction

Assume that you've allotted a US$20,000 budget per person for a one-week boat trip (yacht vacation). You will choose where to travel, which boat to hire and what to do at the places (destinations) you visit. You are required to submit a written report (Term Paper) and present your study in the classroom/online meeting.

Step1: Who is coming?

Decide whom you would like to go on a yacht vacation with. Pick four people (yourself included). They can be real or imaginary people. If your budget is not sufficient for the planned vacation, you may go up to six people, with a corresponding increase in the budget (US$20,000 for each person added). Consider the following:

- What are the names, nationalities (required for visa applications), genders and ages of the people going on vacation?

- Are these people affiliated by any means (family, friend, dating, coworker, etc.)?
- Are there any special circumstances related to these people such as: is vegan, has allergies, disabilities, at an old age or children, coming to this trip for his/her photography hobby, he/she needs to access a cellphone service every two days due to work requirements. You can be creative here.

Step 2: Destination

In this step, carry out the following tasks:

- Select a yacht charter destination far away from home. Figure 17.1 provides a quick snapshot of popular destinations. I recommend picking a destination that is either on another continent or a transoceanic one.
- Decide on the proper time/season for the region you are planning to visit (avoid hurricane season in hurricane prone areas, etc.).
- Gather any further information you may require on the destination you have chosen, including:

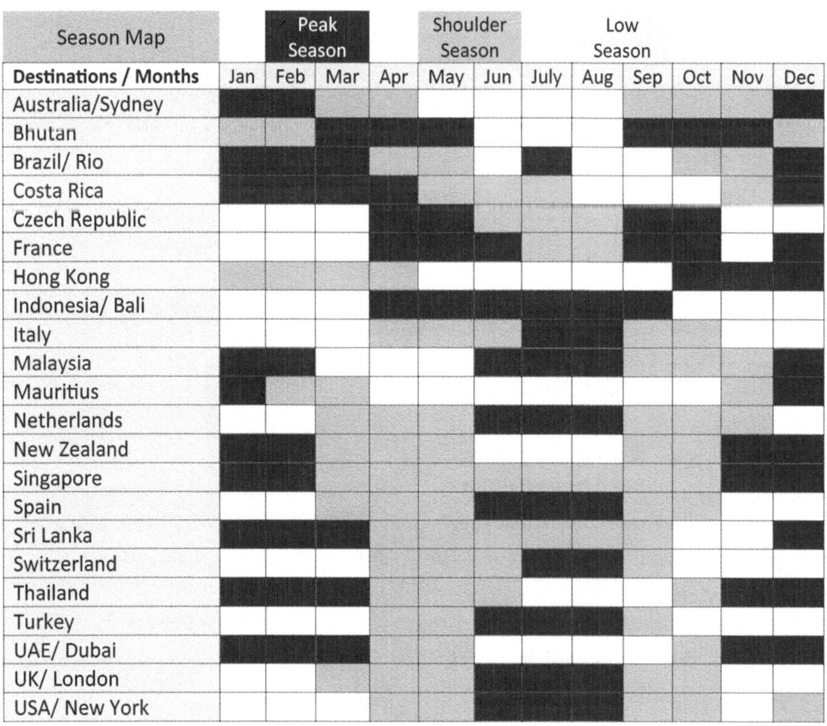

FIGURE 17.1 Quick Snapshot of Popular Destinations

Source: Packer, 2021

  - culture
  - food
  - security
  - touristic attractiveness
  - marinas and airports
  - etc.
- Look for blogs, travel notes and social media entries on the destination that may contain beneficial information for your vacation.

Step 3 Noonsite

Following your decision on which region and countries to visit, you will require additional information about yachting. Almost every yacht traveller (owner/charterer/customer) uses Noonsite to obtain information about new destinations they would like to visit. I would like you to visit the site (www.noonsite.com/countries/) and gather the following information:

- facts such as coronavirus restrictions, time zone(s)
- security, overall crime and safety, maritime security, road security, city security
- weather, oceanographic information (wind, currents, tide, temperature, rain, etc.)
- courtesy flag (You need to hoist a flag of the country visited)
- ports and ports of entry
- formalities
  - general process
  - arrival and departure formalities
  - immigration procedures – visa information
  - customs, restrictions (especially on tobacco and alcohol products), prohibited substances, other rules on the customs
  - biosecurity – covid procedures and documentation
  - documents and declarations
  - fees
- restrictions – restricted areas
- local customs – what to wear on the shore, how to behave, is alcohol consumption is allowed, any local rules that the visitors need to know before landing onshore
- yachting essentials – fresh water supply, fuel, spare parts, repair services
- opening hours and days of shops and businesses
- money/currency used, exchange rates, where to change money
- whether there is an embassy in the country
- emergency phone numbers – police, ambulance
- any necessary information other than the ones asked above.

Browse some of the user comments and reports and share with us those you find interesting/valuable, such as those on piracy, recent changes in bureaucracy or restrictions.

Step 4: The yacht

In this step, consider the following tasks:

- Find a yacht that is in the region or that serves in the region you have chosen by way of Google search or a digital marketing platform such as www.charterworld.com, www.yachtcharterfleet.com,
  - Monohull or multihull? Why?
  - Sail or motoryacht? Why?
- What are the dimensions of the yacht?
- Provide an overview of the ccommodation (number of cabins, facilities, beds and furniture, etc.)
- Provide an overview of the crew onboard.
- Outline the other properties of the boat such as speed, horsepower, bow thruster, stabilisers, air conditioning, etc.
- Carefully examine the list of amenities. Each of the advertised amenities will be discussed and detailed in the project. Some phrases, such as beach club and tender bay, are yacht-specific; pay close attention to those. Try to analyse these features in terms of the yacht's marketing and pricing.
- Write your own subjective evaluation on the yacht in terms of
  - aesthetics
  - colours and decoration
  - interior and exterior design.

Step 5: Activities and fixing the date

In this step, we will be choosing activities to attend and fix the date we wish to be there.

- Determine whether the destination is hosting a timed event/festival/special occurrence. Take a look at how things are marketed/advertised to visiting tourists. Choose whether you want to observe or participate in the activity. Include links to websites and social media screens in your report. Assess the impact of the marketing on the attractiveness of the activity.
- Plan at least a total of *four* activites during your vacation such as:
  - snorkeling – scuba diving
  - swimming
  - visit natural/scenic beauty sites
  - cave, canyon hiking/exploration
  - night life, clubbing, dancing, casino
  - museum, historic places to visit
  - local culture tours, village life
  - temples, churches, mosques, and other religious places of interest
  - surf, sail, SUB, water ski, inflated water toys, and other water sports
  - safari, horseback riding, atv/motorbike/SUV rentals

- shopping, city site seeing
- etc.

If you have decided on the exact date and the activities for your charter, proceed to the next step.

Step 5: Book your flight

Since you have decided to travel to a location far from home, you will need to figure out how to get there. During this phase, you will instinctively Google phrases such as 'plane ticket from Turkey to Fiji' and will most likely obtain some offers. I want you to try out flight-specific platforms and mobile applications like:

- Travelocity
- Expedia
- CheapOair
- TripAdvisor Flights
- Skyscanner
- OneTravel
- Travelzoo
- Google Flights
- Kayak
- Momondo.

Examine various price and transportation possibilities. Allow yourself some leeway – waiting is preferable to missing your connecting flight. REMINDER: You will plan your holiday as one week on the yacht, not total time spent, so it is fine if you include some days either side of the yacht trip.

Once you've decided on a flight itinerary, purchase tickets for everyone who will be joining you on the trip (four if you haven't opted to increase), and include screenshots in the report.

Step 6: Itinerary (voyage plan)

Search for images of 'itinerary', 'travel itinerary' or 'travel itinerary template' on any search engine and choose a format to use, or make your own. Prepare a thorough schedule, going into as much detail as you wish, including timings for 'calling the taxi for the airport,' 'dropping the dogs off with the neighbour,' and 'calling mum to let her know you are fine'.

Optional – If you have the skills of a captain or a deck officer, you may choose to spend time voyage planning and preparing a realistic voyage plan, measuring the course, route legs, distances and timeframes. You may include navigation charts in your report, but keep in mind the yacht's service speed. The distances travelled each day should be reasonable.

Step 7: Day-by-day destination

In this step, you must create a day-by-day travel itinerary. Use visuals of places you've visited/plan to visit, services offered and costs (if applicable), and tell the reader how appealing this destination is and why tourists should visit that area and use the services that are there. The report you generate should have the appearance of a tourism brochure that could be used to promote yacht tourism in the region. In previous studies, some students Photoshopped themselves into scenes of the destinations and shared the experience with the class (snorkeling, dancing with locals, stroking a wild animal, etc.) as if they had truly been there. Students typically choose put a lot of effort into this phase.

Step 8: Food and beverage (Optional)

This step is optional and includes only the food served onboard. Make a catering plan for the intended period:

- Plan all meals including snacks, suppers and special day/celebration gatherings.
- Prepare the menu for each day.
- Plan the purchasing process. Search online for local suppliers, take into consideration if the ingredients are available for the area.
- Budget the food and beverage as part of Step 9.

Pay attention to vegan, kosher, halal and allergies if stated in Step 1.

Step 9: Budgeting

Assuming you had US$20,000 dollars for each individual, create a balance sheet that details each and every expense you have incurred, including:

- flight tickets + additional fees if any applied
- land transportation fees including home to airport, airport to marina travels
- visa and immigration fees
- yacht charter fee + extras paid to yacht company + tip for the crew
- activities, tickets, fees, equipment purchased/rented
- gifts
- all the other expenses you need to consider.

You are not required to spend your entire budget, and you should not feel obligated to spend all of your allocated funds.

Step 10: Conclusion

Using your own words, if you were to sell/advertise this yacht charter vacation:

- What would you highlight?
- How would you attract customers?
- Which media would you market through? Why?
- What would you do differently if you started this study from the beginning?
- How did this study contribute to your learning process?

Reference

Packer, B. (2021). A destination for every season. https://bragpacker.com/a-destination-for-every-season-a-quick-guide-to-picking-the-best-travel-destinations-all-year-long/

Case 18

BOMBARDING THE TOURISM MINISTER WITH QUESTIONS

Gürhan Aktaş and Rut Gomez Sobrino

Duration

Between 90–120 minutes.

Learning objectives

Upon completing the case, participants will be able to:

- explain the importance of crisis communication in tourism
- identify the criteria for organising successful press conferences during crises
- develop skills to organise press conferences and to act as spokespersons.

Target audience

The case aims to emphasise the principles and success criteria of organising a press conference during a crisis and is applicable to all types of tourism management programmes, especially to those where the focus is on equipping students and trainees with managerial skills. In higher education and vocational programmes, the execution of the case, following the completion of courses providing students with background knowledge on tourism systems, tourism marketing, destination management and planning, would result in enabling the students to better assess the potential impacts of the given crisis from the perspective of different subsectors. The application of the case following a lecture on crisis communication strategies would also support the importance of press conferences during crises. However, since participants are expected to learn from their mistakes, instructors/trainers are recommended to discuss the key issues and criteria of a successful press conference following the execution of the case. The case has been executed during crisis

DOI: 10.4324/9781003182856-18

communication training programmes intended for stakeholders by the authors and has proven to be an effective training tool.

The case is designed for a group of 10–15 participants and requires participants to be divided into two teams: the Minister and Press teams. For larger groups, the case can be repeated with different crisis scenarios taking place in different destination settings, and participants can switch teams to engage in discussions concerning both parties. In such cases, the teams formed at the beginning of the case session should be informed on changing roles and on being prepared both as a Minister team for their own crisis scenario, and as a Press team for the press conference presentations of other groups.

Teaching methods and equipment

The case utilises role playing, scenario analysis, group discussion and class discussion as its teaching methods. It is based on media representatives attending a press conference by a Minister of Tourism after a devastating crisis. Therefore, the only need for an effective execution is to arrange a class in theatre-style seating facing a desk where the participant playing the role of the Minister would be seated. Although there is no need for any other equipment, the Minister team can be encouraged to use computers, projectors, flipcharts and/or boards during the press conference based on their availability in the classroom.

Teaching instructions

The following five consecutive steps are explained here in line with the case being executed with a group of ten participants. The steps could be altered based on group size, especially if it is necessary to form more teams for larger groups. Prior to the execution of the case, instructors/trainers can start with an introduction on the importance of crisis management for tourism, and of keeping stakeholders and public up-to-date with information issued by official authorities. It is also worth mentioning during the introduction that crisis communication is not limited to press conferences, but embraces many other tasks and activities.

Step 1

Participants are randomly and equally divided into two groups: The Minister and the Press teams, each consisting of five participants. The groups, seated apart from each other, are then provided with the concerned crisis scenario prepared for each group. The Minister team is asked to discuss how a crisis communication strategy should be formulated, and then to organise a press conference. This team needs to decide which participants will play the roles of the Minister and the moderator during the press conference. This selection should lead the team to discuss the qualities and characteristics of a spokesperson expected from the participant to play the

Minister role. They should also consider potential questions to be directed by the Press team and discuss appropriate answers.

All members of the Press team will play the roles of journalists and news reporters from various media companies. The Press team should distribute the roles of local, national and international media representatives among themselves and should prepare questions to be asked during the press conference. The aim of the Press team is to direct detailed and, to a certain extent, provocative questions to the Minister to get as much information as possible about the crisis. The teams should be given around 15–20 minutes to discuss how they should get prepared for the press conference.

Step 2

The first press conference after the crisis takes place during Step 2. The moderator and the Minister of Tourism selected by the Minister team take their place in front of the class. The moderator commences the meeting by explaining how it will proceed, followed by an introduction of the Minister. The Minister, then, starts with a brief explanation of the crisis and how it is being dealt with. Following the speech of the Minister, a question and answer session managed by the moderator begins. The participants playing the roles of journalists and news reporters should introduce themselves and state which company they represent before asking their questions. Following the question and answer session, the moderator ends the press conference. This step should last around 20–25 minutes.

During Step 2, participants should take notes individually on how this first conference progressed and how the participants played their roles. They should note all positive and negative aspects they have observed during the press conference and should especially assess the attitude, behaviour and responses of the Minister. In this step, participants are likely to make some mistakes. However, the evaluation of the first meeting should be left to the end, when the whole case is being assessed by the participants in Step 5.

The first press conference should be brief and its message should revolve around three main concepts:

1. *Action taking*: The Ministry has immediately called its crisis management team to duty, and assigned it with full authority and responsibility for investigation.
2. *Credibility*: The Ministry is the only official authority to contact for any information regarding the crisis, and the contact details are shared. The schedule of the next press conference will be announced as soon as new information is available.
3. *Sympathy*: The Minister should show sympathy towards those people, whose friends and relatives may have been injured or killed because of the incident, and should assure them that their needs/expectations will receive priority in all actions.

In line with these concepts, some of the common mistakes made by participants include:

- providing too much information and sharing with the press the unconfirmed numbers for the injured and deceased
- blaming other authorities/organisations for failing to prevent the crisis
- making announcements concerning the duties and responsibilities of other authorities without their consent
- the Minister displaying bad temper and acting rude when faced with provocative questions from the press
- selecting a participant who cannot control his/her nervousness, and cannot forward the message clearly, as the Minister
- failing to create an image of a determined, confident, sincere and trustworthy leader, capable of managing his/her crisis management team
- the Minister delivering his explanations more as personal statements rather than the actions taken by a committed and responsible authority
- the Minister neglecting the fact that there are people who are directly affected by the crisis and making jokes and/or answering questions in a joyful manner.

Step 3

In this step, the groups are separated to evaluate and compare their notes on how the press conference went, and how the Minister and the moderator played their roles. The Minister team gets the second part of the crisis scenario with new information, and is asked to get prepared for another press conference. The Press team, not provided with the second part of the scenario, is asked to prepare new questions for the upcoming meeting. The Minister team can either continue with the current participant playing the Minister role, or change to another participant. This would enable both groups to compare the performances of two Ministers. This step should take around 15–20 minutes.

Step 4

The second press conference is staged with the moderator and the Minister taking their place in front of the class. The procedure is similar to that of Step 2, with the only difference being the Minister sharing more information with the Press. The meeting continues with a question and answer session, and is ended by the moderator. All participants should continue with their notes on assessing the meeting and the Minister's credibility and sincerity. This step should last around 20–25 minutes.

During this step, instructors/trainers may aim to detect if the teams pay attention to the following issues:

- According to the imaginary scenario, the second press conference is deliberately scheduled to take place two days after the first one. The Minister team should not make the mistake of accepting that this is the only other meeting they are organising since the start of the incident. They should pretend to have organised at least one more press conference on the second day, since it is important for the press to receive up-to-date information from related authorities during crises.

- For the same purpose, the Minister team is expected to declare that a call centre has been set up immediately after the crisis in order to provide information to all those concerned. In addition, several press releases are expected to have been distributed throughout the crisis.
- The Minister should quote other authorities when providing technical information about the fire, as in with regards to the damaged forest area and the cause of the fire.
- Since the case is based on an imaginary scenario and not all aspects to be considered by the participants can be foreseen, it is very likely that the Press team may come up with interesting and unexpected questions. The participant playing the role of the Minister should respond to such spontaneous questions confidently with logical answers using his/her own creativity and knowledge of tourism.

Step 5

In the final step of the case, participants discuss how they did during role playing. Both positive and negative assessments should be raised, and all participants should be encouraged to contribute to the discussion. Following the discussion, the instructor/trainer can end the case with his/her own assessment and can provide the group with further reading material on crisis communication and press conferences. It has also been found effective to share real-life example videos of successful and rather unsuccessful press conferences. The discussion and conclusion can last around 20–30 minutes.

Some of the key issues that can be utilised by the instructors/trainers to conclude the case study include the following:

- Organisations should pay attention to developing ongoing, consistent and trust-based relations with the media and avoid using media agents only as information disseminators during crises.
- Organising press conferences alone during crises would be a limited approach in crisis communication.
- Organisations should pay attention to disseminating synchronised messages in their communication activities throughout a crisis.
- Crisis communication activities should be realised with supporting crisis management plans that are immediately put into practice.
- The magnitude and scope of a crisis would help organisations in assigning different experts and professionals as the members of a crisis management team for diverse crisis types.

Case

First Part for the Minister team

Dreamland is an island country with a population of 3,2 million citizens. Covering an area of 32,420 sq km, the island's economy highly depends on its tourism industry. Its contribution accounts for the 65% of the total GDP. Last year, the country attracted 5,1 million international visitors, and announced US$6.3 billion tourism income. The country's sandy beaches and clear waters alongside picturesque landscape enriched by forests, canyons and numerous lakes make it an ideal spot for leisure holiday makers, eco-travellers and alternative sports enthusiasts. The leading tourist markets of the country are the USA, Europe, Australasia and Canada. The average duration of stay is six days and the average visitor expenditure is around US$1,850 per person. The Ministry of Tourism, which has been in existence for the last 50 years, aims to develop and promote sustainable tourism and to secure the everlasting benefits of the industry for the country.

On 16th June of this year, which coincides with the high season of tourist arrivals, a forest fire erupted in an inland central region of the country at around 10:00 am, and hastily spread to an area of 145 square kilometres with the help of winds changing directions. In the middle of the fire zone, there is Lake Aquana, home to two 4-star and twelve small-scale hotel establishments. There are also five camping sites, often preferred by trekkers, birdwatchers and other eco-tourists. In all these establishments, it is believed there are around 1,200 tourists and 500 hotel employees at the time of the fire break out. The lake and its surroundings also attract around 200 daily visitors on average in high season. There are around 25 establishments in the form of small markets, souvenir shops, restaurants and cafes catering to these tourists and excursionists. Most of these establishments are run by the locals residing in the nearby two villages, with a total population of 550 citizens.

Although the government has immediately sent firefighting vessels and firefighters to an area, Lake Aquana is unreachable due to the fire blocking the main road. The fire caused a power outage making it more difficult to communicate with the establishments and tourists known to be in the area. The heavy smoke resulting from the fire does not only make it difficult for special firefighting helicopters with water tanks to attack the fire, but also leads to the rerouting of numerous international flights to nearby countries. While the security forces continue their investigations into the cause of the fire, a local news agency issued a report on similar cases caused by campers and tourists previously recorded in the region, but no fire of this magnitude has been recorded in the history of the country. A manager of a travel agency, which organises daily tours to the region, also talked to a local TV station about their tour bus returning from Lake Aquana just a few hours prior to the fire breakout and the bus driver noticing around 40 people on a trekking tour around the place where the fire is believed to have started. Another person claiming to be from one of the villages, said that he managed to talk to his family, who have told him that a German group of eight tourists happens to be missing.

The Minister of Tourism has announced that s/he will organise a press conference at 13:00 today on the recent incident. Prepare the Minister's speech and answers to potential questions in line with the Ministry's crisis management strategy.

First part for the Press team

Dreamland is an island country with a population of 3,2 million citizens. Covering an area of 32,420 sq km, the island's economy highly depends on its tourism industry.

On 16th June of this year, which coincides with the high season of tourist arrivals, a forest fire erupted in an inland central region of the country, and hastily spread to an area of 145 square kilometres with the help of winds changing directions. In the middle of the fire zone, there is Lake Aquana, which is home to two 4-star and twelve small-scale hotel establishments. There are also five camping sites, often preferred by trekkers, birdwatchers and other eco-tourists. The lake and its surroundings also attracts daily visitors on excursions. There are small markets, souvenir shops, restaurants and cafes catering to these tourists and excursionists. Most of these establishments run by the locals residing in the nearby two villages.

Lake Aquana is unreachable due to the fire blocking the main road. The heavy smoke caused by the fire does not only make it difficult for special firefighting helicopters with water tanks to attack the fire, but also resulted in many international flights to be directed towards other nearby countries.

A local news agency issued a report on similar cases caused by campers and tourists previously recorded in the region, but no fire has been recorded on such a big scale in the history of the country. A manager of a travel agency, which organises daily tours to the region, also talked to a local TV station about their tour bus returning from Lake Aquana just few hours before the fire breakout and the bus driver noticing around 40 people on a trekking tour around the place where the fire is believed to have started. Another person claiming to be from one of the villages, said that he managed to talk to his family, who have told him that a German group of eight tourists happens to be missing.

The Minister of Tourism has announced that s/he will organise a press conference at 13:00 today on the recent incident. As the representatives of media companies, get prepared for the meeting.

Second part for the Minister team

Two days later, in the 46th hour of the fire, the Head of the National Fire Department confirmed that the fire has been brought under control. One firefighter suffering a life-threatening injury is in intensive care, while 15 other firefighters and volunteers have suffered minor injuries. The fire burned down 54 square kilometres of forest area, killing hundreds of animals. In the meantime, the Head of the Police Department made another announcement indicating that a piece of broken glass led to the starting of the fire but that it is not possible to track how and when this piece of glass ended up in the restricted forest area. Therefore no person can be declared guilty. The total physical damage of the fire is estimated at around US$1.2 million.

2,057 people were confirmed to be trapped in the region; including 1,080 tourists staying at nearby accommodation establishments, 453 employees and 524 citizens living and working in the region. The early news about 40 trekkers being seen at the spot where the fire started proved to be wrong. There is no evidence of such a group having gone missing. However, the news about a German group has been proven to be true. The group started their trekking tour early in the morning on the 16th June and went missing. The dead bodies of these eight tourists were found on the second day. Their families have been contacted and travel arrangements were made for those who wish to come to the country from Germany. There are also 15 other villagers who have lost their lives, while 57 have suffered various injuries.

The fire threatened the hotels and camp sites at the Lake Aquana region and most of them have been covered in ashes. Although most establishments along the lake side escaped with minor damage, a camping area in the forest was destroyed. All of the visitors staying at the camp site were taken to a safer location by the management in the early hours of the fire. However, two employees who stayed behind lost their lives.

Since daily excursions usually arrive in the region after 11:00 am, the tours organised on the first day of the fire were sent back from their half-way break stop by the police, and no excursionists were harmed.

The Minister of Tourism will hold another press conference on the 17th June at 3.00 pm. Prepare the Minister's speech and answers to potential questions.

Further reading

Aktaş, G., & Gunlu, E. A. (2004). Crisis management in tourist destinations. In W. Theobald (Ed.), *Global Tourism* (3rd ed., pp. 440–457). Burlington: Elsevier.

Glaesser, D. (2006). *Crisis Management in the Tourism Industry* (2nd ed). Abbingdon: Routledge.

Liu-Lastres, B., Schroeder, A., & Pennigton-Gray, L. (2019). Cruise line customers' responses to risk and crisis communication messages: An application of the risk perception attitude framework. *Journal of Travel Research*, 58(5), 849–865.

UN World Tourism Organisation. (2011). *Toolbox for Crisis Communication in Tourism*. Madrid, Spain: World Tourism Organisation.

Zhai, X., Zhong, D., & Luo, Q. (2019). Turn it around in crisis communication: An ABM approach. *Annals of Tourism Research*, 79, 102837.

Case 19

DOOMSDAY IN ŞİRİNCE

Gürhan Aktaş and Burçin Kırlar-Can

Duration

Around 120 minutes.

Learning objectives

Upon completing the case, participants will be able to:

- understand the importance of media relations in destination management and marketing
- describe the effects of positive and negative media coverage on destinations, and their image and carrying-capacity thresholds
- use benchmarking to compare the potential outcomes of extraordinary events in different destination settings
- develop media relations strategies for destinations.

Target audience

This case can be used both in higher education, and vocational and professional training programmes. The case has three parts and consecutive discussion sessions in which participants may associate the events with various disciplines and concepts, ranging from media relations to product development, and from pricing strategies to forecasting methods. Students in the final year of tourism, business or media relations higher education programmes, and professionals with similar previous experience, may attempt to analyse the scenarios from different perspectives, and offer varied opinions on how the destination marketing organisation (Şirince

DOI: 10.4324/9781003182856-19

Tourism Foundation) should have dealt with the doomsday news reports. A lack of extensive tourism knowledge would not cause any problem in implementing the case, as participants should be encouraged to consider and discuss all planning, management and marketing issues. However, instructors and trainers should ensure that the groups reach a consensus on how the organisation should act after considering all potential outcomes of the media coverage. The case is designed to allow participants to reconsider their early reactions to the news, and how their initial decisions changed following new information provided in each part.

The case can be implemented in classes of any size. A duration of 120 minutes is suggested for a class of 25–30 participants, although the duration may vary depending on the class size. The class should be divided into groups of at least five participants at the start. In each group, participants should apportion the roles of the village headman, a hotel owner, a restaurant owner, a wine house owner and a souvenir shop owner, and consider each stakeholder's concerns and expectations when analysing the scenario. After each discussion session, the groups should reach a consensus on how Şirince Tourism Foundation should respond to the news reports to ensure effective planning and marketing regarding the supposed doomsday events across the destination.

Teaching methods and equipment

The teaching methods include role playing, scenario analysis and group discussion. No specific teaching material or equipment is required. The instructors/trainers need to organise the classroom setting to allow the groups of at least five participants to sit together and apart from the other groups, so that each group has some space to engage in discussions and take notes about their discussions.

Teaching instructions

Before providing the teaching instructions, it is important to note that although the case is based on a true story, some of the information provided is not factually accurate but included for educational purposes. For example, there is no official record of the bed and car park capacity in Şirince for the year of this event. Although the figures are logical considering the village's current carrying capacity, they should not be cited for academic purposes. Similarly, since the participants are expected to engage in discussions in consecutive steps without knowing the actual consequences of the news reports, it is important to ask them not to access the internet during the case discussion. Any internet search regarding the news reports and how events developed in 2012 would ruin the group discussions as participants would try to shape their opinions and suggestions according to the actual facts rather than their own ideas.

The three steps are explained here assuming that the case is being executed with a group of 25–30 participants. The steps could be altered according to a group size, especially if more teams are needed for larger groups. Prior to conducting the case, instructors/trainers can start by explaining the importance of representing tourism stakeholders in destination marketing organisations and the need to take destination management and marketing decisions by consensus, reached by evaluating feedback from all stakeholders.

Step 1

The class should first be randomly and equally divided into groups of at least five participants. In each group, participants should share the roles of the village headman, a hotel owner, a restaurant owner, a wine house owner and a souvenir shop owner. After seating the groups apart from each other, they are given the first part of the case scenario. After the instructors/trainers have explained the information provided in this part, the groups discuss how Şirince Tourism Foundation should deal with the news reports and prepare the village for the doomsday event. In this step, participants should be encouraged to consider all potential consequences of media attention on the village, and formulate strategies to guide how the foundation should plan for and manage the expected surge of visitors. The groups should be given around 15–20 minutes for this discussion, after which a selected member from each group should give a five-minute presentation of their suggestions to the class. The instructors/trainers should not provide any feedback on these comments, but can highlight any similarities and differences in how each group explored the scenario and how they think that the foundation should act in response.

Step 2

The second part of the case should be distributed to the groups, who revise their suggestions in light of the new information provided. The groups can be given another 10–15 minutes to decide whether they would like to retain or change their initial suggestions for the foundation's actions. One member from each group then gives a five-minute presentation to the class to explain the reasoning for any changes. The instructors/trainers should only highlight how the groups have responded to the new information, including which aspects of the destination's preparedness they have concentrated on.

Step 3

In the final step, the groups learn about the actual outcomes of the news reports, and the actual number and characteristics of the visitors who travelled to Şirince on the day. The class then considers several discussion questions. Depending on the class size and the time allocated, the instructors/trainers can either ask groups to discuss the questions before presenting their results to the class, or allow the whole class to discuss what they have learned from the Şirince case.

Some of the most important lessons of the case include the following:

- DMOs should develop good relationships with key media agents.
- DMOs should ensure that they are the leading authority to approach for any kind of inquiry about the destination.
- DMOs should employ professional personnel responsible for PR activities.
- DMOs should monitor and manage news reports about the destination.
- DMOs should actively use different marketing tools and channels, such as social media accounts and web sites, to distribute information about the destination.
- DMOs should invest in image marketing activities.
- DMOs should plan, implement and revise marketing actions based on strategic tourism plans.

Case

Part 1

Şirince, a small village located in Turkey with a population of 700, has become a popular day-trip destination, mainly because it is located close to two leading destinations on Turkey's western coast. The first is Selçuk, only a 15-minute drive to the village, which hosts some of the most visited heritage sites in Turkey, including the ancient city of Ephesus, the Virgin Mary's House and the Basilica of St John. The second destination is Kuşadası, which is about an hour's drive to the village. Since the early 1980s, Kuşadası has become an important seaside destination, and strengthened its appeal by building a cruise ship port in the 1990s. Both destinations attract vast numbers of domestic and international visitors. In 2012, for example, when the events in Şirince took place, Ephesus open air museum had 1.8 million recorded visitors (STO, 2016) while 475 cruise ships brought more than 560,000 tourists to Kuşadası Port and around 830,000 overnight tourists stayed overnight in Kuşadası itself (Sezer, 2015; GEKA, 2015).

Şirince not only benefits from its close proximity to these destinations, but it also has village houses with unique nineteenth-century architectural characteristics, the early 18th century St John's Church, a picturesque view of vineyards below the hilltop village, small restaurants specialising in traditional Turkish cuisine and wine houses offering visitors a chance to taste and buy locally produced sweet fruit wines. Most visitors also shop on the village streets from mainly local women selling local products, such as knitted clothes, soaps, olives, locally picked fruits, vegetables and herbs, and house decoration items. There are also small art galleries and specialised souvenir shops selling the work of local artists, such as paintings, and glass and ceramic objects.

Although several village houses have been converted into small-scale accommodation establishments, most international visitors prefer Selçuk or Kuşadası for overnight stays, and come to the village on daily excursions. For those looking for a quiet, peaceful holiday in a natural setting, there are 12 boutique hotels in the centre of Şirince and a camp site within two kilometres, which together have a capacity of around 350 overnight guests. Because Şirince has the narrow streets of a traditional rural settlement, the whole village is pedestrianised, with visitors paying a small daily fee to use one of two car parks at village entrance. The car parks can only hold around 250 automobiles and ten tour buses at a time. Consequently, even on a busy day in the high summer season, the number of visitors does not exceed 1,000 people a day.

In October 2012, Şirince attracted remarkable media attention for an extraordinary reason (Figure 19.1). All Turkey's TV channels and newspapers claimed that the village was one place to remain safe as the end of the world was supposedly approaching soon. According to these reports, the Mayan calendar had predicted that an apocalypse would take place on 21st December 2012, and only Şirince in Turkey and Bugarach in France were safe havens. The reports said that believers in the Mayan calendar and the doomsday prophecy would flock to these two destinations, so tourist traffic would peak. In addition to similar news reports in international media, the number of posts on various social media platforms regarding the prophecy and the village jumped. The consensus seemed to be that around 60,000 domestic and international visitors could visit Şirince on the day.

Discussion

Imagine that there is a small DMO called Şirince Tourism Foundation, and its members include the owners and managers of all tourism-related businesses in the village. Among the member establishments are hotels, restaurants, souvenir shops and wine houses. The village headman also attends the foundation's meetings. Following the news reports, the foundation asks for an urgent meeting with all members to discuss how to deal with the issue, and to prepare for this unprecedented event.

MAYAN APOCALYPSE: TURKISH VILLAGE BECOMES LATEST DOOMSDAY HOTSPOT

A Turkish village has become the latest apocalypse hotspot, with believers of the Mayan calendar prediction that the world will end on December 21 flooding into the area.

FIGURE 19.1 Doomsday in Şirince in the International Media

Source: The Telegraph, 2021

Taking the roles of the village's tourism industry representatives, share your views with your group members.

Part 2

A few weeks after the initial media coverage, other interesting news reports appeared. In particular, an English-language newspaper in Turkey, Daily Sabah, claimed that room bookings in the village had increased for the date of the supposed doomsday, including by the famous Hollywood star, Tom Cruise (Figure 19.2). However, none of the hotels would confirm the sharp increase in bookings, or that Tom Cruise, or any other Hollywood star, had reserved a room.

Another foreign newspaper reported on the other 'safe' village, Bugarach, including an interview with the village's mayor. According to this report, French authorities were more concerned about ritual activities by certain cults than an influx of visitors, since a similar previous event in the Alps in 1995 had resulted in the deaths of a cult's members due to ritual killings. Given that doomsday events had been predicted in the region previously, the mayor called on the local authorities help shut off roads and paths to the village, and manage any potential crowds on the day. The article added that most doomsday prophecies are fabricated by the media and lack any factual basis (Chrisafis, 2012).

In Şirince, several establishments that hoped to benefit from the media attention, came up with interesting and funny product ideas and engaged in small-scale promotional activities to introduce their special 'doomsday' offers. One restaurant, for example, announced its 'doomsday menu' on its social media account. This included

---

TOM CRUISE JOINS THE CROWDS ASCENDING UPON SIRINCE

'Doomsday' arrivals continue to flock to Sirince, a town in Izmir's Selcuk district located on a hill above the Aegean Sea. According to certain Doomsday predictions, Sirince is one of two locations to survive the December 21st end of the Mayan calendar. According to hearsay, amongst the thousands that are expected to take haven in the hillside town is also famous movie star Tom Cruise.

FIGURE 19.2 Daily Sabah News Report on Doomsday Visitors

Source: Daily Sabah, 2012

Soup of the Doomsday, Heaven Kebap, Fire on Rice, Forbidden Apple Desert and Last Breath Tea. A wine producer used a special label, 'Doomsday Wine 2012', on wines to be sold during the day. The producer and his new wine were covered by several national newspapers. The souvenir shops and street stands were also full of t-shirts, mugs, key rings and other objects decorated with doomsday-related slogans and designs. One shop even hired two shop-assistants dressed as an angel and a devil soon after the news emerged in the media. The local regional authority agreed to send its professional dance group to perform an hour-long traditional folkloric show in the main square on 21st December while an events company agreed with one of the hotels to organise a Doomsday party with a DJ performance on the hotel's spacious terrace.

During these preparations, some local people criticised the lack of planning for the anticipated arrival of crowds in a similar way to concerns raised in Bugarach. One stated that a previous cultural festival that attracted around 3,500 visitors a day had resulted in a chaotic environment as the village facilities could not accommodate such large numbers of people simultaneously. He added: 'People could not reach the village because the cars parked on the main road blocked the traffic. The streets were full of garbage. No one bothered to use rubbish bins. Forget about finding a table in restaurants, it felt as if there is not enough space just to stand in the streets. This was with 3,500 people. Now they are talking about 60,000 people. Even if half of them comes, it would definitely be the end of the village, if not of the world'.

Discussion

As the industry representatives, you are invited to a second meeting organised by Şirince Tourism Foundation. In light of the new information provided, discuss how your views on the doomsday have changed since the first meeting. How do you think that the local authorities and tourism establishments should deal with the media attention? What measures need to be taken to safeguard the village from excessive visitors? How should the village prepare itself for the doomsday event?

Part 3

On 21st December 2012, tourism establishments completed their preparations and waited for visitors. Although it was the low tourist season, boutique hotels confirmed that there were no vacancies for that day. Most bookings were by domestic visitors and some media company representatives who travelled to the village to cover the doomsday events. In contrast, there were no bookings by international visitors, and the camp site had received no inquiries or bookings, so it was almost empty.

On the day, nearly 5,000 visitors travelled to the village, although the authorities announced that only 3,500 visitors were present in the village at the same time. These figures were significantly below the expectation of 60,000 people but were still above the numbers the village usually attracted daily both in high season and during special festivals. On-site observations proved that most daily visitors were domestic travellers from nearby towns and cities who had been influenced by the news reports and came to the village for recreational purposes. No visitor groups travelled to the village as members of religious cults or Mayan calendar believers. The doomsday itself turned out to be an entertaining social event during which visitors enjoyed several shows, performances and tourism establishments' special offers.

The next day, the newspaper headlines were all about how the doomsday prophecies were false and that few people had travelled to Şirince. However, it was evident in the village that it had surpassed its capacity for daily visitors. In addition to traffic congestion on the day, the district municipality had to send a team of workers to remove rubbish left by visitors. When asked about the benefits and income generated by the doomsday event, tourism establishments offered contradictory views. The boutique hotels were satisfied with bookings but questioned the manageability of one-day full occupancy in the low season. The restaurants also benefited from the large number of visitors as their income was higher even than the average daily income during high season. There was no clear difference in income generated by those who had developed special doomsday offers and those who had just offered their usual menu items. The difference in return on investment was, however, noticeable in souvenir shops. Those shops that had invested in producing and offering doomsday souvenirs had not achieved high sales as most domestic visitors had bought what the village is known for, such as soaps, knitted clothes and locally picked vegetables, fruits and herbs.

Discussion

- Discuss within your group the lessons learned from the case.
- Do you think that the tourism establishments of the village made mistakes in managing the news reports about the supposed doomsday?
- How would your answers to the questions in the first two parts of the case change following the information provided in this section?
- Do you think that Şirince should consider organising an annual doomsday event?
- How do you think that such an event should be planned, organised and marketed?

References

Daily Sabah. (2012). Tom Cruise joins the crowds ascending upon Sirince. www.dailysabah.com/travel/2012/12/10/tom-cruise-joins-the-crowds-ascending-upon-sirince.

Guney Ege Kalkinma Ajansi (GEKA). (2015). Kuşadası Turizmini Yeniden Pazarlama Stratejisi Projesi: Tanıtım Planı [The Project on Strategies of Remarketing Tourism in Kuşadası: The Plan of Publicity]. Kusadasi.

Selcuk Ticaret Odası (STO). (2016). Turizm Sektör Raporu 2016 [The Industrial Report on Tourism]. Selcuk.

Sezer, I. (2015). Kruvaziyer turizminde dikkat çeken bir nokta: Kuşadası Limani [A striking point in cruise tourism: Kuşadası Port]. *Doğu Cografya Dergisi*, 19 (32), 49–78.
Chrisafis, C. (2012). Bugarach: The French village destined to survive the Mayan Apocalypse. The Guardian, November 19. www.theguardian.com/world/2012/nov/19/bugarach-french-village-survive-mayan-apocalypse.
The Telegraph. (2012). Mayan Apocalypse: Turkish village becomes latest doomsday hotspot. www.telegraph.co.uk/news/worldnews/europe/turkey/9736780/Mayan-apocalypse-Turkish-village-becomes-latest-doomsday-hotspot.html.

Further reading

Fedeli, G. (2020). Fake news meets tourism: A proposed research agenda. *Annals of Tourism Research*, 80, 102684.
McCool, S. F., & Lime, D. W. (2001). Tourism carrying capacity: Tempting fantasy or useful reality? *Journal of Sustainable Tourism*, 9(5), 372–388.

Case 20

NEW MEDIA MARKETING CHANNELS TO MARKET MENGJINGLAI SCENIC AREA

Lanlan Huang

Duration

90–120 minutes.

Learning objectives

Upon completing the case, participants will be able to:

- evaluate the image construction of a tourism destination, and give some suggestions and opinions on how to improve the tourism destination's image
- identify and evaluate the application of new media marketing channels.

Target audience

This case is very representative and typical during learning about new media marketing channels and the image construction of tourism destinations because of the application and development of digital and online technology. As for this case, lots of new media marketing channels have been applied and have ben very effective, which has assisted the destination in attracting a large number of tourists and for its tourism market in Yunnan Province, China to flourish. As such, we can view this case as a successful advertisement activity in new media marketing channels and image construction. In vocational education, this case for students of tourism management would assist in analysing the application of different marketing channels and building a close relationship between tourists and destinations.

We often execute this case in a group of 30–35 students in one class. The instructors can appropriately divide them into teams with a team of four to six persons. Next the instructors will allocate learning tasks to all teams. According to the

DOI: 10.4324/9781003182856-20

learning requirements, every team needs to collect information, data, documents, etc. Following this, the most important point is that they have to analyse what they have collected and share their perspectives, opinions and ideas.

Teaching methods and equipment

Group discussion, brainstorming and task-driven internet searches are applied in this case study. The class should be equipped with computers and multimedia which can improve students' idea sharing. Students in the same group should sit together to discuss and brainstorm.

Teaching instructions

The following five consecutive steps are explained here. Prior to the execution of this case, instructors should do some preparation before the class begins. Firstly, divide the whole class into several groups with a team of four to six persons. Secondly, allocate learning tasks to every group, such as searching information about history, culture, geography and tourism development of the tourist destination. The instructor could encourage students to collect information from different platforms or channels. Meanwhile you could ask students to do self-study about theory and knowledge of new media marketing channels and image construction. Thirdly, ask every group to combine their collections and present them by PowerPoint or any way which can appropriately present their results.

Step 1: A succinct introduction by instructors

Instructors should give a succinct introduction (5–10 minutes) about tourism destination marketing and cases that we will share and analyse in the next stage. The purpose is to ensure students or trainees have the basic knowledge so they are able to concentrate more on sharing and discussion.

Step 2: Information sharing by students or trainees

This step starts with each group being engaged in researching the selected destination, followed by group discussion on findings and summarising these findings for group presentation. The spokesman of each group should present their investigation results including history, culture, geography, tourism development of the tourist destination, and also including theory and knowledge of new media marketing channels and image construction. In this part, each group can present all the factors that have been mentioned above, or they can only introduce one or two factors that they are very interested in. This step lasts around 30–40 minutes. The other participants of each group should make some notes individually and write down differences and discrepancies. They should note all the positive and negative attitudes, behaviours (such as posture, smile, voice, etc.) from every spokesman in order that they can offer some effective and proactive suggestions and advice, and make a fair assessment.

Step 3: Discussion topic one – What is your opinion about image construction?

In the third step, each group will have a short brainstorming around the topic, and also combine their notes on Step 2. They should list the optimistic factors that they can apply in constructing the destination image, and the pessimistic factors that they should do their best to avoid when building a proactive image. Meanwhile, when they share their opinions, it is better for them to combine cases with their results. After finishing their discussions, every spokesman should share their perspectives on behalf of his/her group. When one spokesman finishes their sharing, other participants can cast their doubts or pose questions, then the spokesman or other members in the group should give their response. Attention must be paid to each person being able to freely express themselves, with differences in opinons being respected by all participants. The ultimate aim for participants would be to reach a consensus at the end of discussions. This step should last around 20–25 minutes.

The discussion and presentation should cover three main points as follows:

- how to refine the core features and build the brand image
- the life cycle of a tourism destination and what is beneficial to build the brand image
- the ways and methods of building the brand image of a tourism destination.

Step 4: Discussion topic two – What are your suggestions and advice on applying different new media marketing channels to attract more tourists?

The procedure of this step is similar to the Step 3. The only difference is that participants should focus on arguing new media marketing channels, such as its characters, its pros and cons, its market targets, and so on. Participants should present their arguments in a way to convince other participants with reliable and reasonable explanations. They should also analyse the reason why the same media channels being applied into different destinations would result in diverse results. This step should last around 20–25 minutes.

The discussion and presentation should pay attention to the following issues:

- how to match a celebrity's image with the tourist destinations
- how to select new media channels to advertise the tourist destinations
- how to integrate new media channels in the tourist destination marketing
- how to deal with the relationship between publicity and service.

Step 5: Conclusion

In the last step, we usually do conclusions from the view of participants and the instructor. Each group should make a brief conclusion, including merits and demerits, to give their assessment from their own perspective. Following a discussion, the instructor can share his or her own assessment to wrap up this class, encourage participants to do some further study in their leisure time, and to collect successful or unsuccessful examples or cases to digest theories and promote practices. This step should last around 10–20 minutes.

The instructor could lead the conclusion focusing on the following issues:

- When one successful marketing event has been carried out, some measurements should be taken to keep its reputation in the market and make it sustainable.
- When the attraction of the tourist destination continues to decline, the destination should do some surveys, and then analyse the data to create opportunities to attract more tourists in the future.
- The destination should apply proper new media marketing channels to build effective communication with the tourism customers. We not only pay attention to image publicity, but also value after-sales customer service in order to solidify the good image in the market.

Case

Part 1: Background

Introduction to Mengjinglai Scenic Area in Xishuangbanna Dai Autonomous Prefecture

Xishuangbanna Dai Autonomous Prefecture is located in the southernmost area of Yunnan Province, China. There are 14 tourist attractions, including one 5A tourist attractions, nine 4A tourist attractions, one 3A tourist attraction, two 2A tourist attractions and one 1A tourist attraction. Xishuangbanna governs one county-level city and two counties. There are Jinghong City, Menghai County and Mengla County.

The Mengjinglai Scenic Area in Menghai County is a 4A tourist attraction, developed and managed by Yunnan Golden Peacock Tourism Group Co., Ltd which is a joint venture established by Yunnan Tourism Investment Co., Ltd and Yunnan Mekong Tourism Investment Company. The joint venture mainly engages in the investment, construction, operation and management of tourist attractions in Xishuangbanna. Adhering to the corporate culture of 'mission, loyalty, first-class', the joint venture has continuously improved its core competitiveness and contributed to the development of the tourism industry in Yunnan Province through a strategy of strengthening the development of the tourism industry in Xishuangbanna Dai Autonomous Prefecture.

Tourist attractions managed by the Peacock Tourism Group subsidiary are as follows:

- Xishuangbanna Wild Elephant Valley Scenic Area (national 4A tourist attraction, national nature reserve)
- Xishuangbanna Primeval Forest Park (national 4A tourist attractions, national-level forest parks)
- Xishuangbanna Mengjinglai Scenic Area (national 4A tourist attraction)
- Xishuangbanna Jinuo Mountain Village Scenic Area.

The Mengjinglai Scenic Area is located in Menghai County, Xishuangbanna Prefecture, along with the border between China and Myanmar. It is five kilometers away from Daluo Port on the Kunluo Highway. The clear Daluo River flows through the west side of the stockade, forming a natural border. The other side of the river is Myanmar, so it is known as the 'first village on the China-Myanmar border'.

The Mengjinglai Scenic Area is a typical traditional Dai village. Dai people frequently have contact with and intermarry with the frontiers of the Shan people in Myanmar, forming a rare phenomenon of mixed living between Dai and Shan. The scenic area covers an area of 5.6 square kilometers, and there are more than 100 acres of paddy fields, several fish ponds, idyllic scenery, 58 towers and the thousand-year-old linden trees. The village is natural and elegant. Dai people are warm and hospitable. There are 114 original Dai families in the village. The traditional lifestyle still remains. As for spiritual beliefs and customs, there are Buddhist temples, towers, bodhi trees and ancient springs in the village, which exudes strong cultural characteristics of Theravada Buddhism. Dai people inherit traditional techniques including the old-fashioned paper making, iron forging, pottery making, cloth dyeing, sugar making and wine brewing, called 'living Dai folk handicrafts'.

In February 2010, Mengjinglai was recognised as 'the first batch of 50 characteristic rural tourism villages in Yunnan Province'. In January 2013, it was awarded as a 'China Dai Cultural Protection and Inheritance Demonstration Base' and 'Chinese Cultural Tourism Demonstration Base' by China Cultural Protection Foundation, alongside dozens of other honorary titles.

Main tourist attractions

When you enter the village, you will be embraced by the picturesque view and naturally slow down your pace to enjoy the landscape. You can visit the Entrance Gate of the stockade, the Solemn Pagoda groups, Buddhist temples, Dai traditional culture exhibition hall, villages, 'dry-lane' Dai buildings, boundary monuments and other cultural landscapes. You can visit sacred trees, sacred springs, lovers trees, ancient tree wonders, Peacock River, Border River and other natural landscapes. You can participate in the Water-Sprinkling Festival and other folk cultural festivals. You can experience traditional techniques such as ancient sugar making, slow-wheel pottery,

Dai brocade making, Dai paper making, traditional wine brewing, iron forging and Dai dance.

Multiculturalism

With acres of paddies and fish farms, the Buddhist temple, pagodas, as well as the thousand-year-old sacred trees under the blue sky, Mengjinglai tourist attraction is a paradise far from the crowd. Boasting rich culture, a Buddhist temple in the village is a paradise for pilgrims. It is reported that it is compulsory for local boys to go to the temple in their childhood for cultivating good personality. This primitive but exquisite village retains the distinctive simplicity of the traditional architecture style. Home to hospitable Dai people, you could soak up the traditional vibe and experience Dai people's lifestyle. So it shows a phenomenon of cultural integration, including ethnic minority folk culture, religious culture, agricultural culture and mysterious culture.

Part 2: Introduction to the application of marketing channels

TV show marketing channel

In 2015, when the third series of the Hunan Satellite TV Show called 'Where are you going, Dad' was filmed in Mengjinglai, it attracted brief attention on the internet, which put this small village into the public eye. It also brought a new opportunity for its tourism market. The TV programme 'Where are you going, Dad' is a show that combines travel and education that lots of celebrities and their children participate in. Each of its filming locations become a popular tourist destination, with a market slogan, 'Where dad goes, where tourists go, travel agencies can go'. The programme has brought Mengjinglai a huge number of tourists, and it has also improved the sales of accommodation and tourism-related products, because tourists want to stay one or more nights in village houses or inns. This programme boosted Mengjinglai's popularity in the domestic tourism industry, while it also builds its unique image of rural tourism destination.

New media marketing channels

- *Wechat*: The official WeChat account of Mengjinglai Scenic Area was called 'The First Village on China-Myanmar Border', and now the official account has been changed into 'Mengjinglai Perfume Lotus Paddy Field'. The official account includes entertainment information, accommodation information, online shopping, new events, etc. Tourists can directly contact them. The official account published 61 articles between July 2018 and August 2021.
- *Mini-blog*: Some ads of Mengjinglai Scenic Area have been published on Sina Mini-blog with texts, pictures, videos and links. Many tourists are interested in this scenic area. Many tourists have shared their positive reviews (https://weibo.com/u/5866471947?is_all=1).

- *Douyin*: Mengjinglai opened a Douyin official account which was called 'The First Village on China-Myanmar Border', and now the official account has been changed into 'Mengjinglai Perfume Lotus Paddy Field'. Some short videos were released on this account. The number of positive reviews has reached 11,000. The number of fans is 958.
- *Online Travel Agent*: At present, Mengjinglai Scenic Area has cooperated with Ctrip, Qunar, Xinxin Travel, Qiongyou, Mafengwo, Ali Travel and other platforms in China. In addition, Mengjinglai Scenic Area has joined local tourism e-commerce platforms, offering amazing travelling in Banna, a mobile tour of Yunnan, etc.

Part 3: An introduction to China's tourist attractions quality ratings system

In 1999, the China National Tourism Administration (CNTA) built a national standard named the Standard of Rating for the Quality of Tourist Attractions (National Standard Code: GB/T1775–1999) which could grade tourist attractions from A to AAAA, with 4A being the top grade. Since then, this standard has been revised twice in 2004 and 2012. In 2007, the top grade has been upgraded from 4A to 5A, so currently 5A represents the highest quality of tourist attraction in China. The norm aims to solidify and standardise the management of domestic tourist attractions, to improve service quality, to safeguard the legitimate rights and interests of tourist areas and tourists, and to promote the development, and environmental protection, of tourism resources.

According to the standard of the quality rating system, Local Tourism Quality Evaluation Committees (LTQECs) are authorised and supervised by the National Tourism Quality Evaluation Committee (NTQEC) to accredit 1A to 3A tourist attractions. The National Tourism Quality Evaluation Committee (NTQEC) is authorised by the China National Tourism Administration (CNTA), which was renamed Ministry of Culture and Tourism of the People's Republic of China (MCTC) in 2018. If a tourist attraction wants to be promoted into a high grade, at first they should be accredited a lower grade by local tourism administration and remain at that grade for a few years. As such, 4A tourist attractions rated as 4A for more than three years can apply to national authorities to upgrade to 5A. Time is the basic requirement – during the grading process, the tourist attraction has to face a strict scrutiny, such as key spot check, regular examination, social survey and reviews from tourists. If the scenic spot is not consistent with the specific grade, the application will be rejected or even degraded or delisted. So not only are NTQEC and LTQECs authorised to accredit tourist attractions, but also they are responsible for managing and supervising, and they can check scenic spots in different ways, such as online complaint channels, on-spot inspections, public or private checks, or sampling, etc.

Because of the glamour associated with high-grade attractions, as for local governments, they tend to invest in more high-grade tourist attractions in order to benefit the regional economy through attracting a huge population of tourists and high ticket prices. Because the official standard has a high credibility, it could help the advertising of the attractions, which could naturally push tourists to make the decision to visit them.

Part 4: Discussion topics

1. Mengjinglai tourist attraction is accredited 4A, which is high grade. In addition, it is well-known in our domestic tourism market due to the TV show. Even now when you go through reviews on Ctrip or other online platforms, lots of posts are talking about the TV show with excitement, so what are your opinions about the impact of the official accreditation and the entertainment programme?
2. During the advertisement for Mengjinglai tourist attraction, there are different descriptions of its image construction, such as travelling plus education, the 'First village on China-Myanmar Border' and China's original Dai culture. Which one do you prefer? Which one would attract you at first sight, and why? And what do you think about the function of diverse image descriptions in attracting tourists?

Further reading

Guo, Y. (2017). *Tourism Marketing*. Liaoning, China: Dongbei University of Finance and Economics Press Co. Ltd.

Gao, Y., Su, W., & Zang, L. (2020). Does regional tourism benefit from the official quality rating of tourist attractions? Evidence from China's top-grade tourist attraction accreditations. *Journal of China Tourism Research*, 1–26. doi:10.1080/19388160.2020.1822975

Ge, L. (2017). The First Village on China-Myanmar Border – Mengjinglai. *Creation*, 4, 72–74.

Lin, Y., Lin, B., Chen, M., & Su, C. (2020). 5A tourist attractions and China's regional tourism growth. *Asia Pacific Journal of Tourism Research*, 25(5), 524–540. doi:10.1080/10941665.2020.1741411

Yuan, S., Hao, J., Guan, X., & Xu, H. (2012). The effect of social media on tourism destination marketing: A media-synchronicity-theory based exploration. Proceedings of the International Conference on Services Systems and Services Management, ICSSSM12, *Shanghai, China*, 473–476. doi:10.1109/ICSSSM.2012.6252281

Case 21

TOURIST GUIDES PREPARING TOURISTS FOR SHOPPING ON GUIDED TOURS

Vedat Acar and Abdullah Tanrısevdi

Duration

First step: 30 minutes for theoretical information.
Second step: 30 minutes for each student to read the case studies and answer the discussion questions.
Third step: 30 minutes for group discussions.

Learning objectives

Upon completing the cases, participants will be able to:

- understand the basics of shop visits suiting the needs and expectations of tourists in guided tours
- understand the unethical sales techniques that salespeople apply to tourists in guided tours
- comprehend how disproportionate price reductions affect tourists' trust in guides
- experience how tourist guides lead tourists to shopping on guided tours.

Target audience

Students in tourist guidance departments at faculty and vocational schools.

Teaching methods and equipment

Theoretical expression (tourist guides' duties and responsibilities in guided tours, tourist shops included in guided tour itineraries of travel agencies, preparation of

DOI: 10.4324/9781003182856-21

the tourist groups for shopping by tourist guides), group discussion and souvenir shop visits.

Teaching instructions

Three steps are explained here: theoretical information, three cases and group discussion. The instructor, firstly, gives information about both the duties and responsibilities of tourist guides and the tourist shops visited within the context of guided tours. Then, the cases below are given to students. Upon reading the cases, the students (a maximum of eight students per group depending on the number of the students) discuss each case in the class. Instructors may also include a tourist shop visit (leather, carpet, jewellery, etc.) or take part in a guided tour, which may require obtaining the necessary permissions from the travel agents to observe the genuine shopping environment.

Step 1

The first step starts with the explanation about duties and responsibilities of tourist guides from the beginning to the end of the guided tours. Collecting data before starting the tour, reviewing the timetable, informing the driver, tour staff and the guests about the tour schedule in a detailed way, and allowing the guests to introduce themselves are among those duties and responsibilities. In addition, the instructor also gives information about the tourist shops included in the tour itineraries of travel agencies. Allow 30 minutes for theoretical information.

Step 2

In the second step, in the light of the theoretical information, the students should read the cases and answer the discussion questions.

Step 3

In the final step, the instructor allows the students to discuss each case in the class. The instructor invites students to discuss how the tourist guide prepared guests for shopping. During the discussions, the instructor also asks the groups to reveal possible reasons as to why the expected sales amounts were not reached, how they chould be, and what to pay attention to as a tourist guide in the shopping environment. Thus, in this step, the instructor enables students not only to understand the problems or errors encountered in the cases, but also helps them internalise the critical issues highlighted in the cases. In this way, the students realise that they should not make the same mistakes in their professional life. Allocate 30 minutes for group discussions.

Case I

House of The Virgin Mary-Ephesus Tour, İzmir, Turkey

Friday, 12th August 2016
Cruise Passengers: 26 (12 men, 14 women)

08.40 Most of the group members are 65 years and older. The tour guide starts to talk about Ataturk, the great leader. A few minutes later, the guide gives information about the euro, cent and dollar to Turkish lira rate. Afterwards, she talks about Caravanserai and the meaning of the name Kuşadası. She also emphasises the security of the region despite the negative propaganda in the media.

*09.00*The guide mentions that it is easy to walk in the Ephesus and the tour will come to an end at the Temple of Artemis. Then, she states that the group will finally visit a carpet shop in which hand-made carpets are sold, and adds that no shopping pressure will be applied them there.

12:42 The bus leaves Ephesus. The guide asks, 'How is your tour?' The group members applaud the guide.

*13.24*The guide asks the group whether they would like to visit a carpet shop or not. She also points out that they do not have to buy any carpet in the shop. But there is no response from the group members. Then, the guide adds 'fresh drinks will be served, and the carpet presentation will be performed in an air-conditioned room'. Yet again, there is no response.

13.28 The guide confirms that six people are willing to visit the carpet shop before getting off the tour bus. As soon as the guide finishes her speech, the tourist group gives her a round of applause again. Finally, we come back to the port. Just two Australian tourists out of 26 tourists in the group have been convinced to visit the carpet shop.

There are a lot of carpets demonstrated in the shop. The demonstration takes about an hour. At first, the salesperson does not give any information about the prices of the carpets. After laying the carpets on the floor, the salesperson points out a big carpet whose price is US$4,400. However, the couple does not make any comments about it. Then the salesperson starts to concentrate on alternatives. In the hall, there are just the Australian couple and other employees (the observer, the tour guide and salespeople). A few minutes later, the salesperson lowers the price. As the salesperson is trying to sell the carpet to the couple, the guide puts a piece of paper in one of the couple's hand. Although there is nothing specific about the paper, nor anything written on it, the guide, apparently, wants the salesperson to think they have other plans and need to hurry. Then she points out to the warning above the door stating 'Please, keep door closed as air conditioning is on' and adds that some tourists complain that salespeople force them to buy carpets. According to the tourist guide, the owner of that shop has regarded that warning as necessary.

Our shop visit takes about an hour. Finally, the salesperson asks the couple 'If you had an option to buy, which one would you prefer?' The woman points out a carpet on the floor. Upon this, the salesperson offers US$1,400 for that carpet. After a while, another salesperson says to the other, 'Drop it to US$1,250'. However, the couple states that they have no intention of buying.

14.35 After chatting for a while, they leave the hall.

Discussion

- If you were the tourist guide in this case, what would you do to lead the tourist group to the carpet shop?
- If you were the tourist guide in this case, what would you do to persuade tourists who had no intention to buy?

Case II

German speaking tourist guide with 21 years' experience

Wednesday, 18th January 2017
Morning (09.30–12.30 hours)

I experienced the greatest sadness and tragedy in my life when I guided in a private tour in which two French Canadians participated. They looked like French fashion designers. While visiting Ephesus, I asked them 'What are your occupations?' They replied 'We are fashion designers and live in Paris'. Visualise two men with waist length hair, wearing silk shirts, black trousers, snake leather belts and luxury shoes. At the end of the tour, they sincerely thanked me. Thereafter, we visited a carpet shop. The guests had brought piece of silk with them and seemed as though they were in search of carpet made of silk. The salesperson in the shop offered US$30,000 for a silk carpet without considering the purchasing power of the tourists. I was standing outside the door, right over there. I can assure you that they would have bought because there were pieces of silk in their hands. After hearing the price offered, they told the salesperson 'Thank you, let us evaluate the offered price'. After just a few seconds had passed, the salesperson immediately reduced the price from US$30,000 to US$20,000! US$10,000 discount! Even more, he did not stop and lowered the price to US$11,000! Finally, we left the carpet shop without buying. One of them told me 'Everything is great, you speak English fluently, you are sincere, however, we had trusted you until you brought us here'. That was all! He added 'Please, take us back to the port!' At that moment, I wanted the ground to open up and swallow me!

Discussion

- What criteria should travel agents and tourist guides consider when selecting shops to visit in guided tours?
- How should the travel guide act upon the fact that the salesperson offered a 60% discount?
- If you were the tourist guide, what would you do to regain the trust of the tourists?

Case III

The House of Virgin Mary-Ephesus Tour, İzmir, Turkey

Saturday, 8th October 2016
Cruise Passengers: 8 (7 women, 1 man)

09.35 The guide takes a break in front of the Celcus Library. A few minutes later, the tour leader, arriving with the group from Japan, asks me 'Where is the tourist guide?' Then, I respond that the guide is sitting in the shade right next to the Terrace houses. Upon this, the tour leader goes to her and chats with her for a while.

10.00 We are at the lower gate of Ephesus. The tour guide gives free time to the group. After a while, the tour leader heads for a shop to buy pomegranate juice. The whole of the group follows her and buys pomegranate juices as well. There is no bargaining. Then, the tour guide comes towards me and says 'This is a no shopping tour; however, the tour leader has made a request to visit a leather shop without demonstration.'

10.20 We come to a leather shop in Selçuk, İzmir. As soon as the tourist group gets off the tour bus, the tour leader takes the tourists directly to the toilets in the garden of the shop. From the beginning to the end of the tour, both the tourist guide and the tour leader have talked about toilets. When the group gets back from the toilets, one of the salespeople has greeted them in Japanese in front of the shop. After chatting with the salesperson for a while, the group decides to enter the shop. There are no other tourist groups except for ours.

10.37 A couple goes out of the shop without shopping. The woman wearing a red dress and an old couple in the group are trying on leather jackets. On the other hand, two women are sitting on the sofa in the hall, and they are talking with each other. The salesperson tries to persuade other tourists to buy as well. However, the woman in the white dress looks like she has not been convinced. Then, the salesperson hangs the jacket the woman has tried on on the hanger. But the tourist guide takes the jacket from the hanger, walks towards the tourist, and starts to talk about the jacket.

10.44 The tourist guide has been trying to persuade the woman to buy the jacket. Just a few minutes have passed, the woman tries on the jacket once again and looks in the mirror at herself. Finally, the woman has been convinced to buy.

Discussion

- Considering the primary purpose of the visit is not shopping-related, what is your opinion on encouraging tourists to visit a shop in such guided tours?
- How should the tourist guide organise the shop visit to make sure there are no complaints from tourists?
- What are the responsibilities of travel agents in reference to such shop visits? How should they lead their tourist guides, and deal with tourist complaints?

References

Acar, V. (2018). Investigating international tourists' behaviours: Research on tourists taking part in guided tours in Turkey. *Unpublished Doctoral Dissertation*, Aydın Adnan Menderes University, Turkey.

Acar, V., & Tanrısevdi, A. (2018). Understanding the behaviours of Japanese tourists on guided tours. In M. Kozak & N. Kozak (Eds.), *Tourist Behaviour: An Experiential Perspective* (pp.19–35), Switzerland: Springer, Cham.

Case 22

TANGO OR ACROBATICS

How to capture info groups' memory

H. Kader Şanlıöz-Özgen

Duration

30–40 minutes.

Learning objectives

Upon completion of the case, participants will be able to:

- gauge the feelings of sales agents in info groups at the end of an experiential event
- recognise the impact and outcome of an experiential welcome of info groups on the sales of hotel property
- discuss the impact of surprising sales agents at the info group visits
- create new ways of marketing activities to cater to the needs of sales agents and info groups.

Target audience

Tour operators are still major actors in the tourism industry in generating tourists for destinations. Therefore, it is critical for undergraduate students in the tourism and/or hospitality management programmes to understand how this system works. Participants will benefit from the case more following their courses on the tourism system, tourism/hotel management and tourism/hotel marketing. The case has been formulated to engage up to 20 participants in order to discuss various aspects of such a case in a hotel property and to develop other ways to welcome info groups of sales agents, who are critical for the marketing of hotel properties. The

DOI: 10.4324/9781003182856-22

case is also applicable for a higher number of participants which may create a more fruitful discussion environment in class.

Teaching methods and equipment

The case utilises class discussion and role playing as teaching methods. Depending on the size of the class, rounded seating in a classroom is convenient for the class discussion to strengthen the interaction among participants. A board or flipchart to note the critical terms or concepts during the discussion would be helpful.

Teaching instructions

Participants will be encouraged to share their opinions in the most imaginative and flexible way so that the discussion will flourish. In this way, future managers will be oriented to think outside the box, leading to creative solutions which may be effective in the hotel business and tourism industry.

Class discussion should be organised to find answers to the following questions:

1. What are the experiential aspects used in such events?
   Participants need to recognise and express every single experiential element in the case, including sensory (i.e. music, food and drink products, voice and presentability of the hotel executive, quality of printed materials or gifts), emotional (i.e. fascination, surprise, excitement), cognitive (i.e. information about hotel services), relational (i.e. interactions with the team workers or group members), and behavioural (i.e. testing of hotel services or offerings). They can also analyse the core and supportive services/offerings in a hotel property and link their functionality with experiential aspects so that info group members will differentiate them as distinct features of the property.
2. What are the immediate effects of such events on the sales agents? Please consider emotional, sensory, cognitive and relational aspects.
   Visits from info groups is a typical example of experiential marketing in the hotel business. Participants need to comprehend how this experiential marketing will be effective to capture a place in the memory of travel agents. Therefore, after analysing experiential aspects in the first question, participants will try to assume what immediate outcome can be obtained from those experiential marketing actions.
3. What outcome do you think this hotel had thanks to its special welcome to the info groups?
   Participants need to consider that those groups visit a minimum of seven to eight hotel properties in a day. Therefore, it is quite difficult for them to remember every single detail of a property. Participants need to understand how the hotel management analysed this info group visiting process by prioritising

the experiential side at the visit to the hotel and offering information about the functionality in the printed materials. Participants can also put themselves in the shoes of the info group members and imagine how they would sell the hotels or recommend them to their customers.

4. What other welcome events can be organised for info groups in hotels? Please use your imagination and consider current technological tools to elaborate your opinions.
   Technology has been a major factor affecting marketing activities of businesses including tourism and the hotel industry. Therefore, participants need to take into consideration current technological tools (i.e. AI, Internet of Things, hologram, chatbots, robots, augmented/virtual reality, smart devices) to enhance the experiences of info group members in hotels and to reduce costs of travel to destinations. For example, hotel websites may include a special content for travel agents to find all the answers to frequently asked questions, or an instant messaging tool can be integrated to contact sales agents of the hotels for a specific question or request from a potential customer. Undergraduate students are expected to find many other ways as they are millennials familiar with the dynamics of our world today. Moreover, participants should also think that those visits do not have to be organised in these groups at all times, and the number of people visiting a hotel may change (individual visits for senior agents or managers, groups of 40 people, boutique groups of five to ten people, etc.). Some groups may overnight in the hotel whereas some others may only have a limited time (i.e. 30 minutes to see a hotel property of 25–50 acres).

Case

Antalya is the tourism capital of Turkey, a country among the top ten in the world to attract the highest numbers of international tourists. The destination's tourism is highly dependent on mass tourism. The majority of tourists from major markets book their holidays through tour operators, via online or offline channels, as a package including flight, accommodation and airport transfers. The majority of hotel bookings (80%) come through reservations from tour operators. Moreover, there are more than 200 five-star hotels in the destination which has an extensive location at the Mediterranean rim. Therefore, the majority of hotels feature large scale properties including plenty of food and beverage outlets and facilities to cater the needs of various market groups such as couples, families with children, senior customers, sportive people, etc. The peak season is between June and September, whereas the summer season is defined as being from the beginning of April to the end of October.

The fact that hotels generate the highest level of sales from tour operators means that hotel businesses welcome a number of info groups of sales agents of the travel agencies who sell the products created by tour operators. Therefore, all the hotels in the destination get prepared to welcome info groups as either daily or overnight visits to present their offerings to the sales agents. In general, those groups may include 10–40 sales agents who see at least seven to eight properties a day and stay in one of those hotels at night. Those visits are included in the sales contracts between hotels and tour operators as hotels consider those visits a critical way to improve their sales from those markets.

If a hotel welcomes a group with an overnight stay, this is a chance to ensure the sales agents experience their offerings rather than a quick inspection and observation of the property during a daily visit of maximum half an hour. For this reason, hotel executives try to welcome those overnight groups at a specialty restaurant, at a spa treatment or at a special party so that those sales agents will have beautiful memories of the hotel and recommend it to potential customers, leading to higher numbers of bookings. However, those sales agents as daily visitors visit vast amount of hotels during their three- to four-day info experience and the majority of the hotels they visit look like the same after repetitive tours in the destination.

Knowing this, the guest relations manager of the Lykia Hotel, which was a newly opened property in the destination, opened a discussion at the managerial meeting to ask what kind of memorable event could be organised for such groups. The general manager asked food and beverage and animation managers to think about possible options for a special event to welcome those groups. The hotel had a very creative animation team who suggested a number of splendid shows to present at the hotel property. The choreographer of the team recommended that three shows were presented in the lobby area, which was a huge hall suitable for such events. The selected shows were a tango dance by a couple of dancers, an acrobatic show by three acrobats on a ring and a hip hop dance with three to four dancers. The team also prepared some surprising moments during the tour with pantomime characters who appeared at unusual places to do some activities with the sales agents (i.e. asking them for their ID card, stopping them with short performances before entering a place).

In addition to this live event, the kitchen department prepared homemade lemonade, which is the best drink to refresh sales agents in the summer time. Homemade chocolates were also offered with the drink. All those items were put on tables which were designed as round desks surrounding an animator in a costume. So a magical and surprising atmosphere was created. Tables were moving, dancing or talking when sales agents were taking their drink or chocolate. Animation and food and beverage teams worked under the coordination of the guest relations team who followed up the arrival of the info groups at the hotel property. Once the group arrived at the hotel, the dance show began with a music and sales agents were fascinated by what they experienced. Those events also took place even when the groups came early in the morning and sales agents were even more surprised to see such an event before they started their day. All the info groups from various countries were treated in this way at the beginning of the season. Months later the sales and marketing director thanked the general manager as all the tour operators and sales agents were talking at the tourism fairs and in sales calls about this very special treatment which showed the distinctiveness of the hotel.

Further reading

Gilmore, J. H., & Pine, B. J. (2002). Differentiating hospitality operations via experiences: Why selling services is not enough. *Cornell Hotel and Restaurant Administration Quarterly*, 43(3), 87–96.

KHM Travel Group. (2020). How to get the most out of a FAM trip. YouTube, March 10. www.youtube.com/watch?v=DyaLEITxzvo

Kotler, P. T., Bowen, J. T., & Makens, J. (2016). *Marketing for Hospitality and Tourism*. Upper Saddle River, NJ: Prentice Hall.

Rakhadu Baba. (2021). Fam trip of Sun Hotel & Resort, Abu Road. YouTube, September 16. www.youtube.com/watch?v=q9gSuzIEe34

Schmitt, B. (2011). *Experiential Marketing: How to Get Customers to Sense, Feel, Think, Act, and Relate to Your Company and Brands*. New York: Free Press.

Case 23

I WANT A BRAND NEW CAR

H. Kader Şanlıöz-Özgen

Duration

60–90 minutes.

Learning objectives

Upon completion of the case, participants will be able to:

- analyse the causes of service failure issues and related aspects
- understand the essentials of service recovery strategies and process
- apply complaint management essentials to ensure customer positive response and satisfaction.

Target audience

Service recovery is a critical process for a hotel management to maintain customer satisfaction and loyalty. Therefore, managerial people need to acquire the awareness and skills to analyse service failures and offer relevant solutions and implement necessary recovery actions. Those managerial aspects are essential for undergraduate students in the tourism and/or hospitality management programmes. Participants will benefit from the case more following their courses on services marketing, hospitality operations and security management in hotels. The case is also applicable for participants affiliated within the industry (team workers) in order to improve their managerial skills in order deal with service failure and recovery issues effectively. The case has been formulated to engage up to ten participants in groups of two participants for the role playing part. However, the case is also applicable a higher number of participants.

DOI: 10.4324/9781003182856-23

Teaching methods and equipment

The case utilises class discussion and role playing as teaching methods. Rounded seating in a classroom is convenient for the class discussion to strengthen the interaction among participants. However, some space for the role playing (Step 2) should also be avaialable. A table with a seat on each side would be ideal to create a conversational environment between the customer and the operations manager, and a little area to serve as the lobby where the manager meets the customer. Then, participants can be seated to discuss the service failure and recovery issues.

Teaching instructions

Step 1: Class discussion

All the participants should discuss the following questions:

1. How would you feel if you were this customer or you had a similar incident in a hotel?
   Participants are expected to empathise with the customer. Participants should be encouraged to express feelings and emotions with justifying causes and reasons. Any input from participants will be useful in order to comprehend how such cases may create different feelings among individuals.
2. Which complaint management tactics were applied in the case?
   Participants need to recognise the fact that many complaints in hotels originate from various reasons including maintenance-technical, service related, attitudinal and unusual matters. Participants, therefore, need to ascertain that this is a service-related issue and some training and service procedure actions are required in order to avoid similar cases in the future. Moreover, the customer opts to take complaint action in the hotel, so participants also need to analyse what managerial tactics have been applied in the case (compensatory, managerial intervention, corrective, empathetic or no action).
   The instructor may also open another discussion by asking if the bellboy should be fired from the hotel or punished in another way. This question relates to the managerial competencies of providing a clear job specification for team workers (is car parking the responsibility of a bellboy or a car valet?) and conforming to the current labour regulations in the country. Another discussion may take place to question if the customer should accept the hotel's proposal or pursue a legal case. The discussion should also concern what actions should be taken by the hotel management to overcome this legal claim.
3. How do you think hotel management should manage the relationship with this customer to turn this negative experience into a positive one? Please also consider today's available technological tools to answer this question.
   Social media, especially review sites, is very effective to disseminate positive customer experiences based on the managerial competency of the hotel

businesses. Moreover, customer relationship management (CRM) actions can be integrated to hotel management systems these days. Therefore, a reminder on the system profile of this customer would be helpful for possible future visits of the customer (maybe a VIP remark) so that special attention and treatment will be delivered in order to extend the positive effects of the service recovery process.

Step 2: Role playing

Role playing is effective in developing participants' communication skills and applying their knowledge. They can also be creative and add some more aspects that came out of the class discussion. Role playing should be a voluntary part of this case, but at least one pair of participants should be encouraged. One participant will be the customer and the second participant will role play the operations manager. They will meet at the lobby and the manager will take the customer to the table to discuss the details. The participants will be free to establish their own views and actions to role play the case.

Role playing is critical as participants can experience the situation. They can role play the conditions given in the case but participants may act in a flexible and authentic way during the conversation, including customer reaction, manager response, negotiation and persuasion, and conclusion (the customer may not accept the action of the hotel). After each role play, the audience should discuss if the complaint was handled correctly, what mistakes or missing issues were involved and what else would be applicable to deal with customer reaction. The participant role playing the customer may also ask to the audience for their opinions to accept or not the offer from the hotel management.

Case

It was 2:00 am when the telephone of the operations manager rang at home. She woke up and answered. It was the night manager of the hotel to inform her about an incident with a guest's car. She listened and asked the manager not to move the car until the police came to take the incident statement down. Fortunately, there was no one injured and the case could wait until the next morning without interrupting the guests in the middle of the night. In the morning, when the operations manager arrived at the hotel, she had a short meeting with the night manager to learn all the details of the incident. There was a guest arrival at night. The bellboy drove the hotel car to take the luggage of the arriving guests to the other side of the hotel. When he came back, he did not manage to park the hotel car properly. So the car rolled backwards and crashed into the front of the guest's car.

The operations manager waited some time to ensure that the guests had their breakfast in peace so that she could talk to them calmly. In the meantime, she discussed the matter with the assistant general manager to decide what options were available. After 10:00 am she called the reception to get connected to the guests' room and found out that they were having coffee in the lobby. The operations manager went to explain the situation and invited them to check their car. The guests were unaware of the incident as they had parked their car when they arrived and did not go back later. When they saw the car, which the guest said they bought only a week ago, the gentleman began shouting at the manager. He was so furious that he was asking for a brand new car as this car had lost its value due to that incident. He made some calls to find a lawyer to seek compensation from the hotel management. The operations manager kept silent and waited until the guest had calmed down.

After half an hour, the operations manager asked the gentleman to discuss the subject privately in her office to seek the best solution for them. First she expressed, on behalf of the hotel management, her sadness about this unexpected situation. She explained that the hotel would take care of the repair of the car under the third party insurance of the property. The guests would not be paying anything from his own insurance. The guests were five adults staying in two rooms and they had two more days to stay in the hotel. At this point she offered two solutions to the guest. If the guests needed to return back to their home at the planned departure date, then the hotel would pay all the transportation fees and send the car to them when it got repaired. If the guests were not in hurry to return back, then the hotel would invite them to stay longer until the car was repaired so they could drive it home themselves. This extension was quite hard for the hotel to handle as it was the peak season. However, front office manager was also willing to help as it was his team worker who could not park the hotel car properly.

The guest calmed down quite fast after he saw that the hotel management was helpful and trying to sort out the matter in the best possible way for him. The operation manager also stated that it was the guest's right to go further to open a legal case as all the police reports and incident statements were written. The guest took some time to ask to the other family members. A few minutes later, he came back to say that they had decided to go with the second alternative and wait for the car to be repaired. The operations manager thanked the guest and his family for their cooperation and understanding. She undertook the process of the car repair, followed up the subject personally and informed the guests timely in case of any news. She also coordinated other department managers to take care of any special needs and requests so that the guests would have an enjoyable time to recover from the stress they had experienced due to the incident.

The hotel management needed to prolong the stay only two more days. The car was repaired and the guests left the hotel by thanking the operations manager for her personal care and the hotel management for the professional treatment of the matter. The operations manager and the guests established an emotional bond and the guests continued to come to the hotel in the coming years to enjoy their summer holidays despite the negative experience they had during their first visit.

Further reading

Hoffman, K. D., & Chung, B. G. (1999). Hospitality recovery strategies: Customer preference versus firm use. *Journal of Hospitality & Tourism Research*, 23(1), 71–84.

Wirtz, J., Chew, P., & Lovelock, C. (2017). *Essentials of Services Marketing* (Chapter 13). Singapore: Pearson.

Case 24

DINNER ON THE PIER

A reason to return

H. Kader Şanlıöz-Özgen

Duration

60–90 minutes.

Learning objectives

Upon completion of the case, participants will be able to:

- evaluate the importance and effects of offering special treatment for loyal customers
- recognise the experiential elements that constitute a memorable experience
- formulate a memorable experience including various processes and hotel operations
- discuss the potential outcomes of special treatment offered to certain customer groups.

Target audience

From the suppliers' point of view, the essence of the hospitality experience is to design offerings enhanced by experiential elements so that customers will remember this hotel property and intend to return in future. Therefore, it is crucial for undergraduate students in the tourism and/or hospitality management programmes to understand the importance and implementation of experiential offerings. Participants will benefit from the case more following their courses on tourism/hotel/services marketing and hospitality operations including front of house departments (i.e. rooms division, food and beverage management). The case has been formulated to engage up to 20 participants, split into groups of four

DOI: 10.4324/9781003182856-24

to five participants for the role play. The case is also applicable for participants affiliated within the industry (team workers) in order to improve their managerial and experiential thinking to offer a memorable hospitality experience.

Teaching methods and equipment

The case utilises class discussion and role playing as teaching methods. Rounded seating in a classroom is convenient for the class discussion to strengthen the interaction among participants. However, smaller rounded seating with four to five seats per group will be more practical for communication and interaction during the role play as each participant will role play a relevant department head (manager or supervisor) in the hotel property (kitchen, food and beverage, housekeeping, front office, guest relations – one participant can role play front office and guest relations together in case of four participants). A projector with a computer connection is required for the presentations of action plans at the end of the Step 2.

Teaching instructions

Step 1: Class discussion

All participants should discuss the following questions:

1. How do you think this special dinner was a memorable experience for loyal customers? Please explain, using examples from the case, by analysing sensory, emotional, cognitive, relational and behavioural dimensions.
   Participants need to recognise and express every single experiential element in the case, such as taste of food items in the special menu (sensory), music (sensory), sea breeze (sensory), sea view (sensory), feeling special, recognised, respected (emotional), romantic feelings (emotional), quality of service, food or other items (cognitive), socialisation with other customers and service people (relational), intention to return (behavioural), singing or dancing (behavioural), once in a life experience (behavioural).
   The instructor can also ask participants to imagine what kind of emotions might be evoked as a result of senses that are stimulated (i.e. sight of moonlight to evoke fascination, taste of specific food or drink to create delight).
2. What benefits and outcomes in terms of marketing do you think the hotel obtained thanks to this event?
   Participants should be encouraged to think of immediate and long-term benefits and outcomes. Immediate outcomes might be positive impression, satisfaction, delight of each customer, positive atmosphere shared among customers and hotel team workers. Long-term outcomes should focus on customer loyalty, word of mouth, dissemination of the positive opinions and feelings to other potential customers leading to effects on marketing costs, efficient use of marketing budget for other areas and staff motivation.

3. How do you think hotel management can promote this event using current marketing tools?
   Digital marketing is key for this question and several tools of digital marketing are applicable. For such special occasions, the most impactful channel is social media including TripAdvisor (or other review sites), hotel, customer or team worker social media pages (Facebook, Instagram, etc.) by tagging the hotel property, individuals' names and a special tag created for the event. As most participants will be young people, they can offer creative ways to push customers or team workers to generate social media content.

Step 2: Role playing

Groups of four or five participants will be formed to role play the management/supervisory team in a hotel property. Each participant in a group will role play a department head (manager/supervisor) who will be responsible for organising such an event in a hotel (kitchen, food and beverage, housekeeping, front office, guest relations). Participants can choose their area of interest or expertise in order to indicate the operational details of the plan. They should be seated in a circle to facilitate the interaction, and prepare an action plan outlining all the details of such an event. Each group will prepare a sheet where the workflow and schedule of all departments are defined. The aim is to run the event smoothly without any problem for customers. Participants should only consider the work needed to organise the event. Groups will prepare their action plans in 20–30 minutes. At the end, one representative from each group will present their plan and a group discussion of 5–10 minutes will take place to evaluate the content of the plans by the instructor and other group members.

The action plan can be prepared in a tabular form or work flow chart to include the details to answer Who, What, Where, When, How questions and relevant tasks for each item. The document used for event organisation in hotels is called a function sheet or BEO. This document includes a scheduled list of tasks to fulfil by each department to hold the event. The critical issue is that participants need to consider every single step from the morning until the end of the evening. For example, written announcement on an information board at the beach or information screens in the hotel from the morning may be helpful to inform customers in advance. Housekeeping should prepare and clean the place before and after the event. Food and beverage will deal with seating order as well as sound and light requirements for the music programme. Kitchen should consider the delivery of menu items during dinner. Guest relations can be the master of the event and prepare a dinner programme and list of participating guests so that other departments can plan their operations accordingly. Welcome customers with special treats and music, assistance with their seating and interaction with them to understand how everything is going. A simple memory gift to remember the night should also be considered. If an exclusive menu can be included in the action plan, this will be an extra element to discuss in class in terms of products and their suitability for such an event.

Case

Wonders Hotel was a five-star all-inclusive resort in Antalya, a tourism destination known as the Turkish Riviera. The hotel was among the pioneering properties to introduce the extensive all-inclusive service concept in the destination with a wide range of food and beverage outlets, sports and wellness facilities, and entertainment activities. The hotel recorded a substantial rate of loyal guests who came every year to spend their summer holidays. Hotel management and operation teams managed to create a family atmosphere among the guests and team workers that guest felt like coming home to spend vacation with very good friends. The hotel management used to apply a standard treatment procedure for loyal guests: rooms with sea view or upgrade upon availability, a bottle of wine and fruit basket in the room, and priority for reservations in the specialty restaurants.

However, the general manager was not pleased with this standard procedure for returning guests to appreciate their loyalty to the hotel. Some guests had been coming every year from the opening of the hotel and their children grew up having unforgettable summer memories in the resort. One day at the managerial meeting he shared with operational managers his intention to offer a memorable experience for loyal guests. After a brainstorming between front office, guest relations, and food and beverage managers, the memorable experience was set: A dinner on the Pier. This special fine dining event would be an exclusive experience for only loyal guests to appreciate their loyalty to the hotel. The pier was not used for any other event in the hotel so this would be an exclusive event for loyal guests.

This special event was an experience for the team workers, too, as many preparations and attention to detail were required to hold such an event at the pier. The pier was around 500 metres long, which started at the seashore after an extensive sandy beach of another 500 metres. The dinner took place at the end of the pier right over the sea. The climate is warm and humid in summer so all the equipment needed a sound and careful transfer and maintenance before and after the event. The guest relations manager was assigned as the coordinator of the event which would be organised once a week. Her responsibilities included determining the participant list, invitation to the guests and confirmation of their participation, informing relevant departments about the number of participants and other important remarks, organising the gifts, planning the seating arrangements, welcoming the guests at the event and accompanying them all evening.

Based on the information delivered by the guest relations manager, the housekeeping department was responsible for cleaning the pier; the food and beverage department was responsible for preparing the established menu and setup with banquet tables, chairs and other decoration items; and the sound and light department was responsible for arranging the systems for the live music programme. The dinner started at dusk where the guests were welcomed with a beautiful selection of classical music played usually by a trio. The waiters were specially dressed with fine dining uniforms and the menu was exclusive with quality food products. The live music continued until the end of the event at midnight. Guests were seated with some other guests staying in the hotel, mostly from the same countries. The guest relations manager planned the seating carefully according to their nationalities and professions so that guests who did not know each other would enjoy their time altogether.

The guest relations manager was on duty during the preparations until the end of the night and thus she could have intensive conversations with guests about their experience in the hotel and this special event. All the participating guests were so pleased to have been recognised and valued in such a special way by the hotel management. They shared their feelings about this unforgettable moment which had been such hard work for the relevant team workers. At the end of the night, the guest relations manager invited all the available team workers onto the stage so that guests could applaud them for their extraordinary efforts. One day a guest expressed her feelings to the manager: 'This is once-in-a-lifetime experience and you offer it to us. This is invaluable.' Other guests who were not invited but witnessed the preparations of an extraordinary happening were curious to know what was going on. One day a guest at the beach approached the guest relations manager and learnt what it was all about. The manager was very happy with what she heard: 'We will come back next year to enjoy this unbelievable experience'. This special dinner became a signature event of the hotel which was shared among guests. Some guests were even asking as soon as they arrived at the hotel if they could participate that week's dinner. Hotel management was convinced by the success of this event and continued to hold it for many years.

Further reading

Gilmore, J. H., & Pine, B. J. (2002). Differentiating hospitality operations via experiences: Why selling services is not enough. *Cornell Hotel and Restaurant Administration Quarterly*, 43(3), 87–96.

Kotler, P. T., Bowen, J. T., & Makens, J. (2016). *Marketing for Hospitality and Tourism*. Upper Saddle River, NJ: Prentice Hall.

Schmitt, B. (2011). *Experiential Marketing: How to Get Customers to Sense, Feel, Think, Act, and Relate to Your Company and Brands*. New York: Free Press.

Case 25

DIGITAL COMMUNICATIONS IN TOURISM MARKETING STRATEGIES

Rut Gomez Sobrino

Duration

180 minutes.

Learning objectives

Upon completing the case, participants will be able to:

- identify digital communications applications that can be utilised as marketing tools
- understand the digital tools that can be applied to destinations of different profiles
- develop a digital communications strategy to market destinations.

Target audience

This case study is designed for 15 participants but allows a maximum of 20. These participants should be structured into three groups in the practical component. After the theoretical explanation from the instructor, the three groups will be assigned to develop digital communications strategies for hypothetic destinations that have been challenged by three different types of difficulties: *reputation*, *safety* and *economic turmoil*. These three cases can also be adapted to different types of destinations: *mountain/rural area*, *beach* and *urban*. The groups will receive specific information on the different scenarios proposed.

DOI: 10.4324/9781003182856-25

Teaching methods

As for any practical exercise based on working in groups, the methodology to conduct this case should be based on participation, interaction and exchanges, especially on the digital marketing strategies that the participants should build upon in the different scenarios. The setting of the room in the most appropriate manner to motivate the formation of groups as well as the dynamic debate, together with the adequate technology to facilitate sharing of content and of guidelines, constitute the basic tools and equipment needed to conduct this exercise.

Teaching instructions

The following steps are proposed for the theoretical component that will lead to the practical exercise in groups.

Step 1: Assigning participants the different destination profiles

The three teams should be instructed to work in groups on the types of destinations already identified: mountain/rural destination, beach destination and urban destination.

The following roles are proposed in each team:

- coordinator/speaker of the group
- digital communications strategist
- content producer
- stakeholder manager
- international relations (with global institutions and markets).

Step 2: Instructions to groups

After the instructor makes sure that the participants understand what digital communications is and some key examples of the use of digital communications by Smart Destinations, the three groups will work around the following tasks.

Task1: Define the digital communications strategy

Participants should define the digital communications strategy for the three hypothetical difficulties that each of the destinations (rural/beach/urban) can suffer: *reputation*, *safety* and *economic turmoil*. The strategy should include:

1. *Brand*: what are the distinguishing characteristics that make that destination valuable?
2. *Identity*: specific channels and possible digital communications activities to articulate that brand, for example website, social media platforms, industry networks.

3. *Reputation*: value proposition that builds upon positive stories of travellers, among other stakeholders.

Task 2: Integrate the different activities under the elements of the strategy

Preparation of the strategy, following innovative and creative approaches, should include the different activities under each of the following elements:

- objectives and activities of the strategy
- digital presence in social media and platforms related to the industry
- campaigns on specific aspects related to the destination
- competitions among the user community, stakeholders or other target groups, such as a Media Award.

Task 3: Presentations and exchanges (1 hour)

Presentation of the different strategies developed and exchanges with other proposals. Group presentations should be conducted as follows:

- Each team will present for 15 minutes the digital strategy proposed under each destination profile (45 minutes in total).
- Conclusions from the facilitator of the session will help to polish the guidelines or strategies proposed by the groups (15 minutes).

Case

What is digital communications?

Digital communications encompass those activities related to dissemination, visibility, promotion and engagement that utilise the digital context. Billboards and screens in public places, mobile devices (cell phones and tablets), websites and social media are considered major platforms to articulate digital communications strategies.

A digital communications strategy should be articulated in clear (SMART) objectives, outcomes, activities, messages and channels, and should define implementing partners.These should be structured in a timeline that defines when and how the activities will be conducted together with an evaluation exercise that tells the responsible parties of the implementation if the strategy worked and what areas should be improved.

SMS content when travellers arrive at major attractions, online information connecting all major attractions and services (police, hospitals, museums, etc.),or interviews with key personalities in the destination ,such as the major of the city, are some of the components that digital communications can offer to tourism marketing.

What type of associated tools are developed by the so-called 'Smart Destination Programme'?

These approaches outlined above are part of the considerations included by the so-called 'Smart Destinations' that are defined by SEGITTUR, the Spanish state agency working on Innovation and Technology in the tourism industry as those destinations that (1) utilice ICT to drive competitiveness, (2) enhance efficiency in production and commercialisation and (3) focus on the experience of visitors among other values, such as respecting hosting communities.

Launched in 2019, the Smart Tourism Destination Network (STDN) was proposed by the Secretary of State ofTourism in Spain with the aim to apply digital technologies to sustainable tourism development. The initiative seeks to enhance the experience of the visitor to the destination while respecting hosting communities or residents in that specific destination. Besides proposing a common platform for destinations aligned under the Network, mentoring and training are integrated in the programme.

The ability to walk through the cultural offering of a certain city from a website, to reserve unique attractions from a simple 'click' or to receive a welcome message with travel information when arriving to an airport are some of the resources that digital communications offer to tourism marketing.

At present, the STDN is integrated by 230 members, including local councils of the Group of Heritage of the Humanities Cities of Spain. The value proposition of the STDN has transcended to other regions, like Latin America. The first international destinations –Tequila in Mexico and Medellin in Colombia – are now part of the Network and have observer status.

The Network is now supporting the Future Tourism initiative in Latin America, in cooperation with the Inter-American Development Bank (IADB) by adapting the programme in the region.

Digital communications and their power to market destinations

In present times, innovation and creativity have immense opportunities to develop thanks to the online environment. The digital context poses a wide range of opportunities to market destinations in a unique and effective manner, approaching visitors and other stakeholders, such as investors, in a fruitful and user-friendly manner.

Websites, mapping exercises identifying places of interest and attractions, virtual tours, mobile communications and social media are some of the opportunities that the digital environment offers to tourism marketing.

The structure of a digital communications strategy

Three key areas integrate a digital communications strategy:

1. *Brand*: defining the distinguishing characteristics that make the destination valuable compared to competitors.

2. *Identity*: identifying the specific channels to articulate that brand, for example website, social media platforms and industry networks.
3. *Reputation*: building upon positive stories of users, such as visitors and other stakeholders, especially for start-ups in the tourism industry.

An efficient strategy would be one designing the best activities under each of the areas, ranging from the following (among others):

- Accessible information available on mobile apps that can be advertised at major points of arrival in the destination.
- Websites that include mapping tools to improve the experience of travellers.
- Ad-hoc digital services such as virtual tours around cultural attractions.
- Social media campaigns portraying positive experiences of travellers that can include a contribution from personalities or celebrities.
- Competition on innovative tourism businesses addressing the SDG Agenda.

Further reading

Digital Communications in Tourism. (2016). The more global, the more personal. International Tourism Conference: Promoting Cultural and Heritage Tourism, 2016. www.academia.edu/40634941/DIGITAL_MARKETING_IN_TOURISM_The_More_Global_The_More_Personal

Europen Commission (n.d.) European tourism indicators system for sustainable destination management. https://ec.europa.eu/growth/sectors/tourism/offer/sustainable/indicators_en

Ivars-Baidal, J. A., Celdrán-Bernabeu, M. A., Femenia-Serra, F., Perles-Ribes, J. F., & Giner Sánchez, D. (2021) Measuring the progress of smart destinations: The use of indicators as a management tool. *Journal of Destination Marketing & Management*, 19, 100531. https://doi.org/10.1016/j.jdmm.2020.100531

One Planet Network (2020). Smart Destinations programme. One Planet, Feruary 4. www.oneplanetnetwork.org/initiative/smart-destinations-program

Case 26

MARKETING MIX ANALYSIS FOR DOTE RESTAURANTS IN PORTUGAL

Vahid Ghasemi, Marcelo G. Oliveira and Salar Kuhzady

Duration

90–100 minutes, plus two additional classes for the optional activity (simulation).

Learning objectives

Upon completing the case, participants will be able to:

- understand the main elements related to the marketing mix in hospitality marketing, specifically service-based business
- explain the main components of integrated service management in a local chain restaurant
- analyse local food and beverage marketing and management from a managerial perspective and recommend ideas for growth
- apply critical thinking abilities in new situations.

Target audience

The case study aims to highlight the importance and specificity of the different elements of the marketing mix in the analysis of hospitality business operations. It has been used in undergraduate and graduate tourism marketing and management courses with classes of up to 20 students, divided into groups of three to four members. In this context, it has proved to be a valuable training instrument, namely due to the collaborative and empirical nature of the approach used.

Teaching methods and equipment

Collaborative based learning (CBL): laptop computer or smartphone/tablet. Simulation: desktop or laptop computer, projector.

DOI: 10.4324/9781003182856-26

Teaching instructions

Collaborative based learning (CBL), as a well-studied pedagogical method, is widely suggested as an effective alternative to traditional learning, especially in higher education. In CBL, students are invited to work together in small groups and members are responsible for their own and each other's learning (Laal & Ghodsi, 2012).

Accelerating knowledge construction and deep learning through social interaction, enhancing engagement, critical thinking and communication skills are among the main suggested benefits for learners. In contrast to its opportunities, successfully designing and launching a CBL is not easy. For instance, some students do not have a positive attitude towards CBL because of the lack of meaningful contributions from other members.

Keeping advantages and possible problems in mind, the current case study was based on a collaborative learning experience originally organized for a 'Services Marketing and Management' undergraduate curricular unit.

The implementation of the case study will involve the following steps: (1) pre-class study of the topic, based on documentation provided by the instructor; (2) individual case reading followed by group discussion (in-class session); and (3) presentation of results (in-class session).

First, students are invited to study 'marketing mix in services' a few days before class. For doing so, a learning kit, including articles, books and webpages can be developed and provided to the students.

At the beginning of class, students are divided into teams of three to four members. For each group, three main roles (coordinator, specialist and shaper) are suggested. Groups are then asked to read and discuss the components of the case study, trying to relate the concepts previously studied with Dote Restaurant's marketing and management approach.

As groups will tend to distribute the different components of the case between their members, instructors should point out the relevance of an integrated approach and the need of in-depth discussions considering the interplay of the different elements as an essential step for the development of adequate recommendations.

At the end of the discussion, the following question can be asked regarding each component:

> How can Dote improve its marketing mix in terms of [*component*]?

After group discussions, each team must present their findings in class, including the recommendations for improvement of the different elements of the marketing mix. The session ends with own and peer evaluation.

Possible recommendations that can be provided to the students as examples considering the case studied include:

Product

In order to improve their menu, Dote could offer a greater variety of Portuguese dishes, like the 'Caldo Verde' soup, 'Polvo à Lagareiro' (lagareiro-style octopuss) or

'Arroz de Pato' (duck rice); and in terms of desserts, the famous 'Pastel de Nata' (custard tart).

Place

Given the importance granted to accessibility by the national tourism office, some recommendations in this area could add value to Dote's brand and enhance the social responsibility aspect of the business.

- *Improving entryways*: People with disabilities will naturally appreciate easy access. This means flat entrances, ramps, elevators and automatic doors.
- *Sensory issues*: Ordering a small amount of Braille menus for blind guests can make a big difference, as it allows blind patrons to browse the menu at their own pace, as well as installing sound dampeners to better control the noise levels in the dining room. In the front-of-house hiring process, Dote should keep an eye out for any resumés that list sign language (SL) as one of the applicant's skills.
- *Installing accessible bathrooms*: Consider including one accessible bathroom stall in each restroom. That stall should include a grab bar on the back wall and the side, more space to turn and a wider door that swings out instead of in. Including a sink lower to the ground would also help guests in wheelchairs in particular.
- *Installing baby change facilities in all bathrooms*: Consider installing baby changing stations in all bathrooms, regardless of gender, in order to better accommodate fathers and grandfathers dining out with their toddlers and babies.
- *Reorganising floor layout*: Make sure there is enough space for guests with wheelchairs and other mobility aids to follow the host with ease to their table. The dining room should include special seating to accommodate wheelchairs, including tables that specifically allow enough leg room. Hosts should be trained to pair groups who have mobility issues with that seating.
- *Accommodating service dogs*: Many people with disabilities use service dogs. These animals are legally allowed in restaurants but must always stay with their handler (including when going to the restroom). Because these service dogs are on the job, staff members should resist petting or otherwise distracting the dog unless their handler specifically says otherwise. Offering to bring a water bowl for the service dog is a way to show extra attention to that table's guests.

Promotion

In order to increase the visibility of the restaurant, Dote could broadcast TV spots during the commercial break of programmes watched by the chosen target group. Returning to the idea of marketing to football fans, Dote could broadcast the competition initiative during the break between the first and second half of the derby.

An important part of the communication policy is sales promotion, with which Dote has already reached different audiences, from children (thanks to promotions in cooperation with cinemas) to football fans and also the young, with 'get 2 pay 1' offers.

However, Dote did not take into account online campaigns via e-mail marketing and newsletters, which are very cost-effective and with which a large number of potential customers can be reached.

Process

If Dote wants to expand its target group to include tourists, it must at least take into account the problem of differences in behaviour and habits in different countries. For example, by informing customers that if they intend to eat the appetisers served, these will be added to the bill. It may seem a small thing, but it can save you from bad impressions and negative word of mouth.

Considering the beneficial aspects of experiential learning, namely the promotion of critical thinking abilities, a second, optional activity is suggested, involving simulation as a learning strategy (Hertel & Millis, 2002). The same groups of students are asked to act as consultants and to develop their own case study, choosing a different restaurant/hospitality establishment, conducting fieldwork, analysing its marketing mix and, when necessary, developing adequate recommendations under the supervision of the instructor, namely in intermediate classes. Results can be presented as a written report to the instructor and in pitch form in a formal session to which representatives from the restaurants analysed would be invited, with evaluation taking their feedback into consideration.

Case

The case is meant to serve as an example of the marketing mix analysis students should develop, with the presentation of recommendations when deemed necessary.

Dote restaurant

In October 2015, Paulo Freitas and Henrique Fernandes decided to create a new concept of a beerhouse in one of the most emblematic avenues of Lisbon, Avenida da Liberdade. The first Dote restaurant was then born, its peculiar name formed by the combination of Douro Valley and Tejo River (www.dote.pt). With typical Portuguese cuisine, it was inspired by the traditions of hospitality of the north of Portugal under the management of João Gaito. With a menu in which the *francesinha* is queen and where beer is served all day, there are also cocktails, as well as a great variety of dishes, namely fish dishes for those who want a true Portuguese experience. It is a pleasant and intimate place if you want to go for a drink and enjoy a snack or a quick quality meal. According to reviews on websites such as Yelp,

Zomato, The Fork, TripAdvisor and Facebook, Dote can be generally considered a 'very good Portuguese restaurant'. Although negative opinions are also expressed, the overall feedback and reviews are positive. The average qualification rate never drops below 3.5 stars, and in some websites, it reaches 5 stars, on the edge of excellence. The negative comments refer to different issues, from bad gastronomic experiences to complaints about employees' lack of English proficiency, but they are rather infrequent, especially when compared to the many satisfactory reviews praising the good quality of the food, the friendly and pleasant staff, the rapid service, as well as the comfortable and clean space. Several customers considered it good value for money, and great compliments are made to the protagonist of the place: the *francesinha*.

Analysis of Dote's Marketing Mix

Product

Dote is a Portuguese restaurant located in Lisbon where customers can taste several typical Portuguese dishes. The main product is the *francesinha*, a dish from northern Portugal, more specifically from the Porto district. The *francesinha* is a Portuguese adaptation of the French *croque-monsieur*, originally made with bread, ham, fresh sausage, steak and then covered with melted cheese and a spicy tomato and beer sauce. Ingredients may change, depending on the restaurant or geographical area. Dote has eight different types of *francesinha*, each made to please a different type of customer: the 'Special Francesinha', the traditional one; the 'Dragon Red Francesinha', with extra hot sauce; the 'Pica no Chão Francesinha', a healthier option made with chicken and vegetables; the 'Bitela Francesinha', made with low temperature-cooked calf, mushrooms, roasted peppers, egg and cheese; the 'Vegetarian Francesinha', made with tomato sauce, cheese, vegetables and a vegetarian hamburger; the 'Top Francesinha', the same as the original, but with extra beef; the 'Bifanada', made with Portuguese *bifana* (pork steak), chorizo, sausage, pork cracklings, cheese and mustard; and the 'Bacalhausinha', with crispy dry ham, cod fish, vegetables and shellfish sauce. These are Dote's main products, though not the only ones. Dote also has other typical Portuguese dishes on the menu, like 'Bacalhau à Brás' or 'Prego no Pão', as well as international dishes like hamburgers, BBQ ribs, salads, pastas and steaks, in order to please different market segments.

Dote also offers a kids' menu for those with children (group segment), with grilled steaks, hot dogs and fish fingers. Customers can order traditional snacks and starters like 'Pica-Pau à Dote', smoked ham and cheese and scrambled eggs, for example. To finish a meal, there are several choices: dark chocolate brownie with salted caramel, fresh fruit, homemade chocolate mousse, yogurt and red fruit pie, as well as other desserts that you would normally find in a traditional beerhouse in Lisbon, but with a modern Dote twist.

In terms of drinks, customers can choose from a variety of national and international beers, Portuguese wines from different regions, 'Ginjinha', a Portuguese cherry liquor, Port wine, as well as other alcoholic beverages, water and sodas.

Place

Dote boasts six different physical locations in Greater Lisbon:

1. Dote Alvalade, Lisboa
2. Dote República, Lisboa
3. Dote Odivelas
4. Dote Odivelas Francesinhas
5. Dote Amoreiras, Lisboa
6. Dote Barata Salgueiro, Lisboa

Take-away service is available at all Dote locations.

Online Delivery Service – Dote Online:

1. Av.República
2. Alvalade

Benefits of ordering via Dote online include the option of Family Menu – 24€ and no delivery tax for orders within the Lisbon region.

Delivery Service – Intermediaries:

1. Uber-Eats
2. Glovo

Ordering via intermediary delivery services does not include the Family Menu option and delivery tax is included for all deliveries.

Price

As reported by the TripAdvisor website, prices range from 9€ to 25€ with a quality/price score of 4 stars out of 5. Thanks to the presence of the restaurant in various websites, it is possible to receive promotions if you book through them (The Fork, for example).

Various promotions are advertised on various platforms such as Facebook or Instagram:

- Family Menu with two appetizers, four first courses and two desserts at 58€.
- Menu for two people with one appetizer, two first courses and two desserts at 29€.
- If you order some specific dishes, a second portion is offered.

- 'Nomes da Semana', a promotion in which clients with the chosen name of the week are entitled to a 50% discount.

Menu price analysis

- 'Entradas': the section of appetisers consists of 11 dishes with an average price of 5.70€. It starts from a minimum price of 0.65€ for the 'rissol' to a maximum of 12€ for the 'pata negra'.
- 'Saladas': this part presents four suggestions at an average price of 10.87€ starting from the 'Cesar' at 9.95€ to the 'Tropical' at 10.50€.
- 'Do Porto': this is the section of the menu with six types of 'francesinhas' (the house specialty) and two types of hamburgers. The minimum price for a 'francesinha' is 11.95€ and the highest is 12.95€, while the cost of the hamburgers ranges from 11.50€ to 12€.
- 'Dote Mini': this is a very clever part of the menu which offers several dishes in a reduced size. There are seven mini versions, with an average price of 7,50€.
- 'Carnes': ten different types and cuts of meat are proposed, the least expensive at 10€ and the most expensive at 19€. The average price for this section is 14.30€.
- 'Peixes': seven fish dishes with an average price of 12.80€.
- 'Sobremesas': Nine desserts are offered at an average price of 3.50€.

On Tripadvisor, every time the price is mentioned in reviews it is always accompanied by positive remarks as people consider the price excellent considering what they have eaten.

Promotion

Drawing customers' attention to the product, the restaurant has its own website (www.dote.pt) and is present in various online platforms, such as TripAdvisor, The Fork, Zoomato, Yelp, Untappd, Visitlisboa.com, Google and Foursquare. Dote, therefore, can be said to have a strong online presence.

Although Dote is a large chain, the target group seems to be quite limited, as the prices are not suitable, for example, for teenagers. However, thanks to offers such as the children's menu or the lunch menu, Dote can cater to different target groups such as families with children, but also businessmen and couples. A very important initiative has been launched by Dote regarding the sports industry, particularly football. Thanks to this initiative, on Dote's website, customers can try to win an official Porto, Benfica or Sporting jersey (the main football clubs in Portugal). For this reason, it can be said that Dote has opted to broaden its target clientele based not only on the age of the customer, but also on sporting interests.

In terms of social media presence, promotion must be designed in such a way that it reaches potential customers and is positively received. In addition to the preferences of the target group, the choice of communication platform is also decisive. Dote is present on Facebook, Instagram, and, more recently, on YouTube. Dote uses a number of offline and online advertising channels:

- *Offline advertising measures*: Dote is not present on TV, newspapers or magazines, for example. On the other hand, they were recently featured on a Portuguese radio programme on 'Rádio Comercial'.
- *Online advertising*: Dote has video advertising on YouTube and Facebook. Although the site is well maintained, there is a noticeable lack of display marketing or display advertising.

Public relations

Dote recently managed to come to an agreement with three main football teams in Portugal in order to provide a positive image of the company through a competition that allows customers to enter the code received on the receipt for a chance to win an official jersey from one of the three teams.

Word of mouth

Dote has gained a large percentage of its customers by word of mouth. Word of mouth is very important in the digital information age, when one has limited time to process large amounts of information. Satisfied customers will recommend products and services, talking about them in a 'genuine', objective, unsolicited way. Therefore, it is important to build good relationships with customers, so that they can always speak positively about the venue and about the products.

People

Customers

The prices in Dote target certain types of clientele:

- families
- groups of friends (over the age of 18)
- tourists
- businessmen
- couples.

The segment is large enough to form a profitable market.

Staff

The staff is mainly composed of the dining room, kitchen and cashier staff. It is very important to understand the importance of professional training, as a positive customer experience will always be the key to increasing results in the restaurant business. In the restaurant, each person has his/her own task. In addition, empathy and politeness are particularly important. The 'people' element of the 7Ps, however, involves, directly or indirectly, anyone who works on the business side of the enterprise. Therefore, the importance of people also extends 'behind the scenes' i.e., to every person who, through their work, helps to create and disseminate the positive image of the restaurant. For example, all the people involved in branding, marketing in general and in everything related to customer support/service.

Process

In order to understand how the process works, you need to go inside the restaurant and understand how the staff welcome the customer. Above all, you need to understand the level of customer care during the customer experience. The staff working at Dote is well divided into cashiers, kitchen staff and waiters. Given this, it is easy to understand that the restaurant responds positively in terms of speed of service. The courtesy and empathy shown by the staff throughout the restaurant experience is also very important. The website offers the possibility to book a table or even to buy the product online. And it is present in the delivery apps. This is very important, because it reduces work and increases profit, as the process is very simple and intuitive online, with offers suitable for everyone. The online process also considers the importance of the customer-seller relationship, providing all the necessary information to contact the restaurant.

Covid-19 has caused problems in all economic parties, including the restaurant industry. Dote has responded positively to this problem, thanks to technological innovation, eliminating menus and replacing them with digital ones that can be scanned with a mobile phone using QR codes. What is more, at the end of lunch or dinner, staff disinfect the tables immediately. Thanks to technology, Dote has dealt with the loss of customers through the restaurant's presence on both Glovo and other delivery apps.

In conclusion, process is something that Dote addresses very well. The restaurant has a standard appearance when compared to restaurants of its type, it does not have its own customisation. That is, it should have features that make the staff unique, for example, knowing how to advise on what combinations to make with the chosen food. In addition to matching, cross-selling would be a good strategy, and it is very important especially for those who go to the restaurant industry for the first time. This is because cross-selling means offering the customer one or more additional products which they had not planned to buy. This would lead to an increase in the revenue of the restaurant, or at least in the shopping cart for online orders. We would like to mention that, in general, Dote can also be classified as a tourist restaurant. Through a small analysis of the reviews on the major social networks such as Google, Trip Advisor and the Fork, we noticed a topic that prevails: the 'entradas'. For example, in Italy it is usual to offer small entrées such as ham or bread. Usually, but almost always, these starters are offered by the restaurant and are intended to give you a taste of something while you wait for the dishes to be served. Some reviews, however, report that Dote serves appetisers that have not been requested by customers whicg are are then added to the final bill.

Physical evidence (Dote República, Lisboa)

Atmosphere

The restaurant has a good atmosphere. It has a very rustic vibe because of its wooden design, which energises the space. It has a river wallpaper associated with the name 'Dote' itself. Dote uses small yellow light bulbs, with its task lighting the hanging décor around them in most corners and areas. However, it also uses small, white light bulbs with an inverted cone around them at the counter area and the bar beside it. The combination of both lights gives a cozy feeling for the customers while they wait for their food, eat their meal, have a chat, etc. The restaurant is huge, so it can be a little noisy, especially if it is full of people. Its music is mixed, focusing on local and modern songs. The scent is positive and refreshing, but it is the aroma of food that can usually be sensed.

Layout

The restaurant is divided into four sections. The first section, outside, is occupied by five wooden tables and four wooden chairs for each table. Another section, inside, near the counter, has a high, long wooden table and 14 high leisure chairs. Another interior section displays one short couch chair at each table and a military green leather long sofa attached to the wall, which can be divided for people to sit on. The last section has four wooden tables and two-seater sofas for each table around the corners, usually occupied by four people. Dote Restaurant has a spacious, cozy and functional layout. Even though the restaurant is divided into four sections, it has a similar design concept for all of them. It has artificial plants, mainly on the corners and parts of the ceiling, which display an 'artsy style'. Its impressive abstract design on the floor and wooden pattern captivates people. Its colour scheme includes a brown wooden colour, shades of blue, yellow, green and dirty white, specifically for the ceiling. The front range is where employees take customer orders and where employees give the customer's order to the chef.

Artifacts

Dote Restaurants do not have a lot of artifacts, apart from portraits of rivers, *francesinhas* and other dishes. The abstract tiles on the floors and the brick wooden pattern add to the idea that it is inspired by a historic or ancient theme, aided by one of the images, that of the iconic metal bridge 'Luís I' crossing the river Douro in Porto, in the north of Portugal.

Equipment

Cutlery is used in Dote restaurants. As the menu includes typical dishes of Portuguese cuisine, forks and knives are provided. On each table, there are napkins, cutlery, yellow or plain glasses for water (or other drinks) and paper placemats. Regarding the menu, if you are sitting indoors, they will give you the menu and write down your orders on paper. It has televisions and air conditioning.

Signage/Logos

In this restaurant, signage and logos are clear, guiding customers through the facility smoothly.

Website

Dote Restaurant's website provides adequate information about the restaurant. It includes its recent offerings, pictures of the area, foods and other associated apps (which are associated with the restaurants for delivery).

Building

Dote restaurants are either located inside a mall or a building in local areas. The interiors of the restaurants are similar, while the exterior design is different. Some entry areas of the restaurants contain a barrel design, the menu in a standee and a 'Super Bock' 90s design, while others only have the big 'DOTE' logo with bright white lights in it.

Uniform

The staff's uniforms are plain black shirts with the 'DOTE' logo. On some occasions, however, near Christmas, Valentine's Day or Easter, for example, members of staff are dressed in long sleeved striped shirts.

Packaging

Dote restaurant provides and uses aluminum packaging, which protects and preserves food products. It also uses carton packages for some types of food.

References

Hertel, J. P., & Millis, B. J. (2002). *Using Simulations to Promote Learning in Higher Education.* Sterling: Stylus Publishing.

Laal, M., & Ghodsi, S.M. (2012). Benefits of collaborative learning. *Procedia - Social and Behavioral Sciences*, 31, 486–490.

Further reading

Lovelock, C. H., & Wirtz, J. (2004). *Services Marketing: People, Technology, Strategy* (5th ed.). Upper Saddle River, NJ: Pearson.

Wilson, A., Zeithaml, V. A., Bitner, M. J., & Gremler, D. D. (2012). *Services Marketing: Integrating Customer Focus across the Firm* (5th ed.). Boston, MA: McGraw Hill.

Case 27

WHO REALLY MAKES THE HOLIDAY DECISION

Is it the buyer or the seller?

Demet Ceylan and Erhan Bilgici

Duration

90–120 minutes.

Learning objectives

Upon completion of the case, participants will be able to:

- identify the maximum impact of guest profiles in holiday decision making
- evaluate guests' target sale potential based on their previous stay and comment cards (survey), as well as their current expectations via social media, customer relations management, etc.
- establish a marketing approach that offers guests the personal touch by fulfilling a need.

Target audience

This case emphasises the principles for recovery following the sudden cancellation of reservations under the constraints of tour operator mass sales and marketing conducted for all-inclusive resort destinations. The marketing of holiday products and services is the result of a joint effort between establishing the value of the holiday and the expectations of the customer. Participants should be informed about the all-inclusive 3S (sun, sea, sand) holiday concept and the domination of tour operators in mass tourism vs individual Guest Touch villas. This case is applicable to all students, from undergraduate to Ph.D. Group work is recommended for role play, but individual brainstorming can also be carried out.

DOI: 10.4324/9781003182856-27

Teaching methods and equipment

The case uses role playing, brainstorming and class discussion as teaching methods. Role play or brainstorming groups can be arranged into teams, such as hotel management team, GEM team, villa management team and guests. A classroom with grouped tables is adequate since the GEM and the customer relationship management (CRM) database and telephone calls to guests will be made via imaginary role play.

Teaching instructions

This course can be conducted as role play, brainstorming or a class discussion, depending on student numbers.

Step 1

Organise students into the following groups:

1. *Hotel executives*: The executives are under stress and believe the solution to their problems lies with the tour operators. To reach high occupancy levels, the hotel must discount its prices. There is a reluctance to approach free individual travellers (FITs). In this case, it is usually the instructor who creates the stress factor.
2. *CRM and GEM team*: Role play consisting of unorganised data, good quality data in revenue per head (REVPAR), occupancy rates, duration of stay, reservation channel (tour operator and individuals), payment channel (who made the payment) and good quality data for contacts with reference to email and telephone number.
3. *Villa team*: They are willing to take the lead and are prepared to approach past guests. Photos from last season's welcome cocktail party are left at villa doorsteps (indicating the number of guests, their names and reservation dates and duration of stay). Two previous team members remember the guests and their personal preferences. It is easy to find the extra costs for special event celebrations during the stay from the butler's archive.
4. *Guests*: Alternative scenarios.
   a. Special days are approaching as reasons for a celebrative holiday, such as:
      a family member's birthday is approaching
      a wedding anniversary is approaching
      a religious holiday is approaching.
   b. Reasons boosting the need for an escape holiday include:
      - parents are working as CEO/CFO and have limited free time to spend with their two children.
      - a family member is recovering from a serious injury/illness (the villa team learns about this from Facebook posts)
      - husband proposed marriage at the same hotel several years before.

(The instructor may give a hint to the role player to encourage positive or negative reactions to make it more convincing).

Common mistakes made by participants at this stage include:

- failing to gather information on past stays, e.g., party size, length of stay, period of stay, complaints/compliments, guest preferences and limitations, and budget spend during their last stay
- failing to recognise department boundaries within the organisational structure and the tensions that can be created due to the penetration of other departments' jurisdiction.
- generating arguments that can bring the villas team to the forefront and leave the others behind, and portraying the villas team members as the conquerors
- assuming an understanding of guest values based on participants' past experiences and failing to discuss the core values and key concepts or touch upon these values with team members.

Step 2

In the second step, groups should write down their arguments, considering possible reactions of other groups, and collect ideas for approaching potential target groups. Students should bring role play into action in the following ways:

1. The villa team need to convince the hotel executives that there is an appropriate way of approaching past guests for sales opportunities and overcome the opposition of the sales team who claim that the villa team cannot generate sales.
   Hotel executives have three reaction options:
   a. *Positive reaction*: supporting the villa team.
   b. *Reluctant reaction*: listening to the villa team's plan but with reservations about the reputation of the hotel, the balance between departments, and not being sure about the impact of contacting past guests.
   c. *Negative reaction*: not interested in supporting the villa team's plan.
2. The villa team needs to cooperate with the CRM and GEM team to collect relevant data and organise it logically, and gather further data from guests' Facebook and social media profiles.
   The CRM and GEM team have three reaction options:
   a. *Positive reaction*: supporting the villa team.
   b. *Reluctant reaction*: listening to villa team's plan but with reservations about losing their area of responsibility, and not being sure about the future impacts of such interference.
   c. *Negative reaction*: angry at the villa team for breaching their area of responsibility.

3. Villa team approach guests (role playing different service solutions that will appeal to and convince the decision maker in the family).
4. Guests are approached by the villa team offering a holiday package. The villa team assumes that they are fulfilling a need or offering a solution based on the above presented customer scenarios.
   Guests have three reaction options:
   a. *Positive reaction*: welcoming the offer, tending to agree.
   b. *Reluctant reaction*: welcoming the offer but with reservations, second thoughts, need for approval from husband/wife, etc.
   c. *Negative reaction*: not interested in the offer.

During Step 2, participants should take individual notes on the way communication is conducted, which strategies are utilised, and how well the participants played their roles. They should note all positive and negative aspects observed during the meeting and, in particular, they should assess the attitude, behaviour and responses of the groups. In this step, participants are likely to make mistakes. However, Step 1 should not be evaluated until the whole case has been assessed by participants in Step 5.

Step 2 role play sessions should be brief, and the content should revolve around three main concepts:

1. *Taking action*: cooperating with and convincing hotel executives/CRM and GEM teams/villa team/guests with teamwork leading to revenue generation by selling holiday packages to past villa guests.
2. *Credibility*: obtaining past records from the hotel database for understanding the past preferences of guests, establishing a business plan with the potential number of guests to approach and potential revenue generation, confirming current posts from social media and contacting hotel guests.
3. *Empathy*: communicating with guests based on their needs and values based on the above scenarios and proposing solutions via tourism products.

Common mistakes made by participants at this stage include:

- delivering too much information about the holiday product instead of addressing the values/needs of the guest, affecting their sincerity and empathy
- failing to check that the family status is up-to-date, e.g., divorce, new baby or death, and entering into a conversation without testing the current status of the family can also cause loss of empathy
- failing to prepare answers to questions like, 'How much did we pay last time?' or 'When was the last time we stayed at your hotel?', leading to a loss of credibility.
- pushing for sales rather than listening to the guest's current or changing values, needs and expectations

- blaming other departments for not acting efficiently or for bad record keeping
- creating forceful entry to their department's jurisdiction (i.e., proposing prices without prior approval from sales and marketing or GM)
- questioning the authority of GM and other department managers publicly.

Step 3

In this step, the groups are split up to evaluate and compare notes to ascertain where role play was successful and to recommend areas for improvement, such as credibility of available data and addressing the values of guests with empathy. This step should take approximately 15–20 minutes.

Some common mistakes made by participants at this stage are as follows:

- failing to empathise with the other participants
- failing to keep the tone of comments constructive
- voicing excuses rather than listening to feedback.

Step 4

Similar to Step 2, participants conduct another role play to:

- convince the hotel executives
- cooperate with the CRM and GEM team
- approach guests.

Each participant should note whether improvements have made been made since Step 2 and recommend areas for development.

During this step, instructors/trainers may suggest paying attention to the following issues:

- hotel management's attitudes while under pressure from last-minute cancellations
- GEM and CRM teams' reaction to the villa team while under the impression that the villa team is attempting to take over the role of sales and marketing
- reactions of guests based on their values and the attitude of the villa team.

Step 5

In the final step of this case, participants will discuss their performance during role play. Both positive and negative assessments should be raised, and all participants should be encouraged to contribute to the discussion. Following the discussion, the instructor/trainer can conclude the case with their own assessment and provide the group with further reading material on tourism products with special restrictions on all-inclusive 3S tourism and value-based communication. The discussion and conclusion should last approximately 15–20 minutes.

The instructor is advised to take videos of the role plays, to be given to each of the participants. This will allow the participants the privacy they need for self-reflection.

Case

Setting

With over 300 days of sunshine a year, 640 kilometres of coastline and beautiful beaches, Antalya is one of the major destinations located on the Mediterranean coast of Turkey and catering for sun, sea and sand. The main procurement agents for holiday resorts with an average of 500–600 rooms are tour operators. The half-a-million-bed capacity of Antalya is mainly sold in bulk as packages tours to the Russian, German, UK and Ukrainian markets. The domestic market is the second priority with the exception of during two major religious holidays per year. Sales and marketing teams focus on inbound tourism bulk buyers, as Free Individual Traveller (FIT) sales channels are not yet fully developed.

GEM systems, also known as CRM, are integrated with tour operator reservation systems for mass tourism, whereas the FIT database requires manual data entry that can create inconsistencies or poor data quality. Sales and marketing teams focus on international source market activities rather than domestic market brand creations or communication. More than 80% of occupancy is between 1st March and 31st October as the majority of hotels only operate during the summer season. Allotment contracts with tour operators cover the summer season and price adjustments during the season are made via special offers. A pricing strategy for FITs is usually not developed.

There are a limited number of resorts with villas for guests who are willing to pay premium prices for tailor-made services and private facilities such as a pool, private beach, private dinner arrangements, private excursions and a butler service for individual organisations. The number of villas is limited in luxury resorts where the per night price of a villa is six times more than a standard double room with a sea view. Villa guests are high revenue generators with extra services like special day celebrations, and they spend far more compared to regular room paying guests. Privacy and a personalised approach are the main reasons for the guests to prefer villas.

In November 2015, all allotment contracts for the 2016 season were signed and sales and marketing teams were confident of an increase in sales; however, during the 2016 season the downing of a Russian aircraft by the Turkish authorities increased political tension between Russia and Turkey. The incident led to the political decision to ban all commercial flights to Turkey, which rendered all allotment contracts for the Russian source market null and void. This dissipated 35% of the season's turnover due to a 65% cancellation in villa reservations and a 30% cancellation in main building reservations.

Scenario

Top management and sales and marketing departments in a resort are under pressure to generate revenues. Therefore, to reach the optimum occupancy level for the summer season they are discussing alternative approaches including:

- discounting prices for existing tour operators since they are already operating direct flights to Antalya
- finding new source markets and tour operators and establishing flight routes via İstanbul (loss of flight comfort at an increased package price)
- finding new source markets without visa implications (gap between reservation and arrival)
- introducing additional all-inclusive services (all food, beverage and entertainment services are already included in the room price).

Are there alternative strategies that can be generated?

The villas manager, being the hardest hit by this political issue, has tried to discover and implement an untravelled road to top management. The villa team could approach previous guests and entice them with a tailor-made service based on their past holiday preferences. The villa team needs support from the GEM team to merge the available CRM online data and photo archive data, as well as updating guest profiles from publicly available sources, such as social media. In addition, the villa team have requested support from the sales and marketing team on pricing and potential discounts and/or addition of services free of charge to adjust the invitation offer per guest.

For the past two seasons, villa guests have been welcomed by the villa team as if they were arriving on the doorstep of their own private mansion. As per Turkish hospitality, arriving guests are welcomed with a warm smile and offered refreshment to help them relax after their journey. Villa guests are offered champagne and snacks, flowers are given to the female guests, there is a live violin performance and a photographer takes pictures of the guests in natural poses during welcome cocktails. As a surprise, the very same night, while the guests are having dinner in the à la carte restaurant, a special table is arranged for them. It contains printed, framed photos from the welcome meeting. These photos are placed in the guests' rooms to create a warm homely feel. The villa team is small but very talented. The team consists of two members with personal knowledge of their guests, which adds a personal touch.

Challenges for the villa team include:

- identifying the villa guests with a high potential to purchase
- identifying the decision maker in the family

- approaching the guests with a special offer to match their values and expectations
- cooperating with the GEM on data mining and alignment to avoid offending guests.
- cooperating with sales and marketing teams while experiencing a challenging pricing environment that applies to all hotels in the destination.

Are there any other challenges the villa team should consider?

Short-term target

Two major holidays for Turkey are approaching and the domestic market appears to be more vibrant during these holidays. Can the villa team generate revenues from domestic repeat guests?

Mid-term target

The Ukrainian market shares similarities with the Russian source market in terms of luxury holidays. Can the villa team cooperate with the sales and marketing team to approach VIP tour operators from Ukraine?

Long-term target

As villas are quite different compared with the hotel rooms where most of the tourists are staying, can the sales and marketing team cooperate differently with the villa team to assure long term profitability and occupancy of villas from different source markets with exclusive services and products? How can the GEM and CRM teams support and develop a suitable database for the villas in which to keep specific data for family preferences, values and expectations, and to be used for marketing strategies?

Further reading

Kozak, M. (2010). Holiday taking decisions: The role of spouses. *Tourism Management*, 31(4), 489–494. https://doi.org/10.1016/j.tourman.2010.01.014

Kozak, M., & Karadag, L. (2012). Who influences aspects of family decision making? *International Journal of Culture, Tourism and Hospitality Research*, 6(1), 8–20. https://doi.org/10.1108/17506181211206216

Ozdemir, B., Çizel, B., & Bato Cizel, R. (2012). Satisfaction with all-inclusive tourism resorts: The effects of satisfaction with destination and destination loyalty. *International Journal of Hospitality & Tourism Administration*, 13(2), 109–130. https://doi.org/10.1080/15256480.2012.669313

INDEX